D0984759

Synchronous and Resonant DC/DC Conversion Technology, Energy Factor, and Mathematical Modeling

Synchronous and Resonant DC/DC Conversion Technology, Energy Factor, and Mathematical Modeling

Fang Lin Luo

Nanyang Technological University
Singapore

Hong Ye

Nanyang Technological University
Singapore

A CRC title, part of the Taylor & Francis imprint, a member of the Taylor & Francis Group, the academic division of T&F Informa plc.

This material was previously published in *Advanced DC/DC Converters* © CRC Press, LLC 2003.

Published in 2006 by
CRC Press
Taylor & Francis Group
6000 Broken Sound Parkway NW, Suite 300
Boca Raton, FL 33487-2742

Printed in the United States of America on acid-free paper
10 9 8 7 6 5 4 3 2 1

International Standard Book Number-10: 0-8493-7237-2 (Hardcover)
International Standard Book Number-13: 978-0-8493-7237-7 (Hardcover)
Library of Congress Card Number 2005050887

Library of Congress Cataloging-in-Publication Data

Luo, Fang Lin.
Synchronous and resonant DC/DC conversion technology, energy factor, and mathematical modeling / Fang Lin Luo, Hong Ye.
p. cm.
Includes bibliographical references and index.
ISBN 0-8493-7237-2
1. DC-to-DC converters. I. Ye, Hong, 1973- II. Title.

TK7872.C8L86 2005
621.381'044--dc22 2005050887

Taylor & Francis Group
is the Academic Division of T&F Informa plc.

Visit the Taylor & Francis Web site at
http://www.taylorandfrancis.com

and the CRC Press Web site at
http://www.crcpress.com

Preface

DC/DC converters are systematically sorted into six generations. We introduce DC/DC converters of the fifth and sixth generations in this book. The purpose of this book is to provide synchronous and resonant DC/DC conversion technology, energy factor and mathematical modeling that are both concise and useful for engineering students and practicing professionals. It is organized into 240 pages and 100 diagrams to introduce more than 50 topologies of synchronous and resonant DC/DC converters. All prototypes are novel approaches and great contributions in modern power engineering. In addition, energy factor and mathematical modeling for power DC/DC converters are introduced. We have systematically sorted the contents into four groups:

1. The fifth generation (synchronous) converters;
2. The sixth generation (multiple-element resonant power) converters;
3. DC power sources, control circuits, EMI/EMC, and application examples;
4. Energy factor and mathematical modeling for power DC/DC converters;

The fifth generation (synchronous) converters have a large number of prototypes published in the literature. We introduce five types of the updated synchronous converters to summarize this technique. From these examples, people can understand the clue to designing synchronous converters.

The sixth generation (multiple-element resonant power) converters involve the two-element resonant power converters (2E-RPC), three-element resonant power converters (3E-RPC), and four-element resonant power converters (4E-RPC). There are eight topologies of 2E-RPC, 38 topologies of 3E-RPC and 98 topologies of 4E-RPC (2C-2L). We carefully discussed one 2E-RPC, one 3E-RPC and two topologies of 4E-RPC in this book to summarize this technique. From these examples, people can understand the clue to designing multiple-element resonant power converters.

We introduce various DC power sources and control circuits for DC/DC converters. In addition, we carefully discussed the EMI/EMC problems in power DC/DC converters. Some successful applications are listed in the book for people to understand the importance of power DC/DC converters nowadays.

The DC/DC converters' characteristics in steady state have been successfully discussed in literature. We still have to investigate the transient process of power DC/DC converters. Mathematical modeling for power DC/DC converters is a historical problem accompanying the development of DC/DC conversion technology. Many experts have tried their best to model DC/DC converters since the 1940s, but no sufficient model was established.

Energy storage in power DC/DC converters has received attention in the past. Unfortunately, there is no clear concept to describe the phenomena and reveal the relationship between the stored energy and the characteristics of power DC/DC converters. We have theoretically defined a new concept — energy factor (EF) — and researched the relations between EF and the mathematical modeling of power DC/DC converters. EF is a new concept in power electronics and conversion technology, which thoroughly differs from the traditional concepts, such as power factor (PF), power transfer efficiency (η), total harmonic distortion (THD), and ripple factor (RF). EF and subsequential other parameters can illustrate the system stability, reference response (unit-step response), and interference recovery (impulse response). This investigation is very helpful for system designing and foreseeing DC/DC converters' characteristics.

This book is organized into eight chapters. Chapter 1 introduces synchronous converters; Chapters 2 to 5 introduce multiple-element resonant power converters. Chapters 6 and 7 introduce DC power sources, control circuits, EMI/EMC, and application examples; and Chapter 8 introduces energy factor and mathematical modeling for power DC/DC converters.

Our acknowledgments go to the executive editor of this book.

Dr. Fang Lin Luo and Dr. Hong Ye

Nanyang Technological University

Singapore

Authors

Dr. Fang Lin Luo received his Bachelor Sc. degree, first class with honors, in radio-electronic physics at the Sichuan University, Chengdu, Sichuan, China, and his Ph.D. degree in electrical engineering and computer science at Cambridge University, UK in 1986.

Dr. Luo was with the Chinese Automation Research Institute of Metallurgy (CARIM), Beijing, China as a senior engineer after his graduation from Sichuan University. He was with the Enterprises Saunier Duval in Paris, France as a project engineer in 1981–1982. He was with Hocking NDT Ltd., Allen-Bradley IAP Ltd., and Simplatroll Ltd. in England as a senior engineer in 1986–1995 after he received his Ph.D. degree from Cambridge University. He was with the School of Electrical and Electronic Engineering, Nanyang Technological University (NTU), Singapore from 1995.

Dr. Luo is a senior member of IEEE. He has published 7 teaching textbooks and 218 technical papers in *IEEE Transactions, IEE Proceedings* and other international journals, and various international conferences. His present research interests are digital power electronics and DC & AC motor drives with computerized artificial intelligent control (AIC) and digital signal processing (DSP), and DC/AC inverters, AC/DC rectifiers, and AC/AC & DC/DC converters.

Dr. Luo was the chief editor of the international journal, *Power Supply Technologies and Applications*, from 1998 to 2003. He is the international editor of *Advanced Technology of Electrical Engineering and Energy.* He is currently the associate editor of the *IEEE Transactions on Power Electronics* and *IEEE Transactions on Industrial Electronics.*

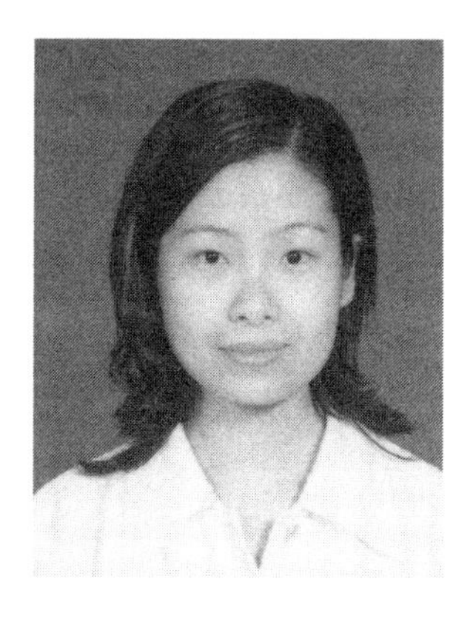

Dr. Hong Ye received her Bachelor's degree (the first class with honors) in 1995 and the Master Engineering degree from Xi'an Jiaotong University, China in 1999. She completed her Ph.D. degree in information technology and structural biology (IT & SB) at Nanyang Technological University (NTU), Singapore.

Dr. Ye was with the R&D Institute, XIYI Company, Ltd., Xi'an, China as a research engineer from 1995 to 1997. She is with NTU from 2003. Dr. Ye is an IEEE Member and has co-authored seven books. She has published more than 48 technical papers in *IEEE Transactions, IEE Proceedings* and other international journals, and various international conferences. Her research interests are DC/DC converters, signal processing, operations research, and structural biology.

Contents

1 Synchronous Rectifier DC/DC Converters 1
1.1 Introduction 2
1.2 Flat Transformer Synchronous Rectifier Luo-Converter 5
1.2.1 Transformer Is in Magnetizing Process 5
1.2.2 Switching-On 6
1.2.3 Transformer Is in Demagnetizing Process 6
1.2.4 Switching-Off 6
1.2.5 Summary 7
1.3 Active Clamped Synchronous Rectifier Luo-Converter 7
1.3.1 Transformer Is in Magnetizing 8
1.3.2 Switching-On 8
1.3.3 Transformer Is in Demagnetizing 8
1.3.4 Switching-Off 9
1.3.5 Summary 9
1.4 Double Current Synchronous Rectifier Luo-Converter 9
1.4.1 Transformer Is in Magnetizing 10
1.4.2 Switching-On 11
1.4.3 Transformer Is in Demagnetizing 11
1.4.4 Switching-Off 11
1.4.5 Summary 11
1.5 Zero-Current-Switching Synchronous Rectifier Luo-Converter 12
1.5.1 Transformer Is in Magnetizing 13
1.5.2 Resonant Period 13
1.5.3 Transformer Is in Demagnetizing 13
1.5.4 Switching-Off 14
1.5.5 Summary 14
1.6 Zero-Voltage-Switching Synchronous Rectifier Luo-Converter 14
1.6.1 Transformer Is in Magnetizing 15
1.6.2 Resonant Period 16
1.6.3 Transformer Is in Demagnetizing 16
1.6.4 Switching-Off 16
1.6.5 Summary 16
Bibliography 17

2 Multiple Energy-Storage Element Resonant Power Converters 19
2.1 Introduction 19

2.1.1 Two-Element RPC ... 20
2.1.2 Three-Element RPC ... 21
2.1.3 Four-Element RPC ... 22
2.2 Bipolar Current and Voltage Source ... 24
2.2.1 Bipolar Voltage Source ... 26
2.2.1.1 Two Voltage Source Circuit ... 26
2.2.1.2 One Voltage Source Circuit ... 27
2.2.2 Bipolar Current Source ... 29
2.2.2.1 Two Voltage Source Circuit ... 30
2.2.2.2 One Voltage Source Circuit ... 30
2.3 A Two-Element RPC Analysis ... 31
2.3.1 Input Impedance ... 31
2.3.2 Current Transfer Gain ... 32
2.3.3 Operation Analysis ... 33
2.3.4 Simulation Results ... 37
2.3.5 Experimental Results ... 38
Bibliography ... 38

3 Π-CLL Current Source Resonant Inverter ... 41
3.1 Introduction ... 41
3.1.1 Pump Circuits ... 41
3.1.2 Current Source ... 41
3.1.3 Resonant Circuit ... 42
3.1.4 Load ... 42
3.1.5 Summary ... 42
3.2 Mathematic Analysis ... 43
3.2.1 Input Impedance ... 43
3.2.2 Components' Voltages and Currents ... 44
3.2.3 Simplified Impedance and Current Gain ... 45
3.2.4 Power Transfer Efficiency ... 52
3.3 Simulation Results ... 53
3.4 Discussion ... 54
3.4.1 Function of the Π-CLL Circuit ... 54
3.4.2 Applying Frequency to this Π-CLL CSRI. ... 55
3.4.3 Explanation of $g > 1$ ... 55
3.4.4 DC Current Component Remaining ... 55
3.4.5 Efficiency ... 55
Bibliography ... 55

4 Cascade Double Γ-CL Current Source Resonant Inverter ... 57
4.1 Introduction ... 57
4.2 Mathematic Analysis ... 57
4.2.1 Input Impedance ... 58
4.2.2 Components, Voltages, and Currents ... 59
4.2.3 Simplified Impedance and Current Gain ... 60

4.2.4 Power Transfer Efficiency 66
4.3 Simulation Result 67
4.3.1 $\beta = 1, f = 33.9$ kHz, $T = 29.5$ µs 69
4.3.2 $\beta = 1.4142, f = 48.0$ kHz, $T = 20.83$ µs 69
4.3.3 $\beta = 1.59, f = 54$ kHz, $T = 18.52$ µs 70
4.4 Experimental Result 71
4.5 Discussion 73
4.5.1 Function of the Double Γ-CL Circuit 73
4.5.2 Applying Frequency to This Double Γ-CL CSRI 73
4.5.3 Explanation of $g > 1$ 73
Bibliography 73

5 Cascade Reverse Double Γ-LC Resonant Power Converter 75
5.1 Introduction 75
5.2 Steady-State Analysis of Cascade Reverse Double Γ-LC RPC 76
5.2.1 Topology and Circuit Description 76
5.2.2 Classical Analysis on AC Side 77
5.2.2.1 Basic Operating Principles 77
5.2.2.2 Equivalent Load Resistance 77
5.2.2.3 Equivalent AC Circuit and Transfer Functions 78
5.2.2.4 Analysis of Voltage Transfer Gain and the Input Impedance 80
5.2.3 Simulation and Experimental Results 84
5.2.3.1 Simulation Studies 85
5.2.3.2 Experimental Results 86
5.3 Resonance Operation and Modeling 86
5.3.1 Operating Principle, Operating Modes, and Equivalent Circuits 87
5.3.2 State-Space Analysis 89
5.4 Small-Signal Modeling of Cascade Reverse Double Γ-LC RPC 92
5.4.1 Small-Signal Modeling 93
5.4.1.1 Model Diagram 93
5.4.1.2 Nonlinear State Equation 93
5.4.1.3 Harmonic Approximation 94
5.4.1.4 Extended Describing Function 95
5.4.1.5 Harmonic Balance 96
5.4.1.6 Perturbation and Linearization 97
5.4.1.7 Equivalent Circuit Model 98
5.4.2 Closed-Loop System Design 99
5.5 Discussion 104
5.5.1 Characteristics of Variable-Parameter Resonant Converter 105
5.5.2 Discontinuous Conduction Mode (DCM) 108
Bibliography 114
Appendix: Parameters Used in Small-Signal Modeling 116

6 **DC Energy Sources for DC/DC Converters** 117
6.1 Introduction 117
6.2 Single-Phase Half-Wave Diode Rectifier 118
6.2.1 Resistive Load 118
6.2.2 Inductive Load 119
6.2.3 Pure Inductive Load 122
6.2.4 Back EMF Plus Resistor Load 123
6.2.5 Back EMF Plus Inductor Load 125
6.3 Single-Phase Bridge Diode Rectifier 125
6.3.1 Resistive Load 127
6.3.2 Back EMF Load 129
6.3.3 Capacitive Load 131
6.4 Three-Phase Half-Bridge Diode Rectifier 133
6.4.1 Resistive Load 133
6.4.2 Back EMF Load ($0.5\sqrt{2}V_{in} < E < \sqrt{2}V_{in}$) 134
6.4.3 Back EMF Load ($E < 0.5\sqrt{2}V_{in}$) 136
6.5 Three-Phase Full-Bridge Diode Rectifier with Resistive Load 136
6.6 Thyristor Rectifiers 138
6.6.1 Single-Phase Half-Wave Rectifier with Resistive Load 139
6.6.2 Single-Phase Half-Wave Thyristor Rectifier with Inductive Load 140
6.6.3 Single-Phase Half-Wave Thyristor Rectifier with Pure Inductive Load 141
6.6.4 Single-Phase Half-Wave Rectifier with Back EMF Plus Resistive Load 142
6.6.5 Single-Phase Half-Wave Rectifier with Back EMF Plus Inductive Load 144
6.6.6 Single-Phase Half-Wave Rectifier with Back EMF Plus Pure Inductor 145
6.6.7 Single-Phase Full-Wave Semicontrolled Rectifier with Inductive Load 147
6.6.8 Single-Phase Full-Controlled Rectifier with Inductive Load 148
6.6.9 Three-Phase Half-Wave Rectifier with Resistive Load 149
6.6.10 Three-Phase Half-Wave Thyristor Rectifier with Inductive Load 151
6.6.11 Three-Phase Full-Wave Thyristor Rectifier with Resistive Load 152
6.6.12 Three-Phase Full-Wave Thyristor Rectifier with Inductive Load 153
Bibliography 155

7 **Control Circuit: EMI and Application Examples of DC/DC Converters** 157
7.1 Introduction 157

7.2 Luo-Resonator 157
7.2.1 Circuit Explanation 158
7.2.2 Calculation Formulae 159
7.2.3 A Design Example 160
7.2.4 Discussion 160
7.3 EMI, EMS, and EMC 161
7.3.1 EMI/EMC Analysis 161
7.3.2 Comparison to Hard-Switching and Soft-Switching 163
7.3.3 Measuring Method and Results 163
7.3.4 Designing Rule to Minimize EMI/EMC 167
7.4 Some DC/DC Converter Applications 168
7.4.1 A 5000 V Insulation Test Bench 168
7.4.2 MIT 42/14 V 3 KW DC/DC Converter 169
7.4.3 IBM 1.8 V/200 A Power Supply 171
Bibliography 173

8 Energy Factor (EF) and Mathematical Modeling for Power DC/DC Converters 175
8.1 Introduction 175
8.2 Pumping Energy (PE) 177
8.2.1 Energy Quantization 177
8.2.2 Energy Quantization Function 177
8.3 Stored Energy (SE) 177
8.3.1 Stored Energy in Continuous Conduction Mode (CCM) 178
8.3.1.1 Stored Energy (SE) 178
8.3.1.2 Capacitor-Inductor Stored Energy Ratio (CIR) 178
8.3.1.3 Energy Losses (EL) 179
8.3.1.4 Stored Energy Variation on Inductors and Capacitors (VE) 179
8.3.2 Stored Energy in Discontinuous Conduction Mode (DCM) 180
8.4 Energy Factor (EF) 182
8.5 Variation Energy Factor (EF_V) 183
8.6 Time Constant t and Damping Time Constant τ_d 183
8.6.1 Time Constant t 183
8.6.2 Damping Time Constant τ_d 184
8.6.3 Time Constants Ratio ξ 184
8.6.4 Mathematical Modeling for Power DC/DC Converters 185
8.7 Examples of Applications 186
8.7.1 A Buck Converter in CCM 186
8.7.1.1 Buck Converter without Energy Losses ($r_L = 0\ \Omega$) 186
8.7.1.2 Buck Converter with Small Energy Losses ($r_L = 1.5\ \Omega$) 190
8.7.1.3 Buck Converter with Energy Losses ($r_L = 4.5\ \Omega$) 192

8.7.1.4 Buck Converter with Large Energy Losses ($r_L = 6\ \Omega$) 196
8.7.2 A Super-Lift Luo-Converter in CCM 198
8.7.3 A Boost Converter in CCM (No Power Losses) 201
8.7.4 A Buck-Boost Converter in CCM (No Power Losses) 206
8.7.5 Positive Output Luo-Converter in CCM (No Power Losses) 209
8.8 Small Signal Analysis 211
8.8.1 A Buck Converter in CCM without Energy Losses ($r_L = 0$) 214
8.8.2 Buck-Converter with Small Energy Losses ($r_L = 1.5\ \Omega$) 215
8.8.3 Super-Lift Luo-Converter with Energy Losses ($r_L = 0.12\ \Omega$) 218
Bibliography 223
Appendix A: A Second-Order Transfer Function 225
A1 Very Small Damping Time Constant 225
A2 Small Damping Time Constant 226
A3 Critical Damping Time Constant 228
A4 Large Damping Time Constant 228
Appendix B: Some Calculation Formulas Derivations 231
B1 Transfer Function of Buck Converter 231
B2 Transfer Function of Super-Lift Luo-Converter 231
B3 Simplified Transfer Function of Super-Lift Luo-Converter 232
B4 Time Constants τ and τ_d, and Ratio ξ 232

Index 235

Figure Credits

The following figures were reprinted from *Power Electronics Handbook* (M.H. Rashid, ed.), Chapter 17, Copyright 2001, with kind permission from Elsevier.

Chapter 1:
Figure 1.5
Figure 1.6
Figure 1.7
Figure 1.8

Chapter 7:
Figure 7.1
Figure 7.2

1

Synchronous Rectifier DC/DC Converters

The global merchant dollar market for DC/DC converters is projected to increase from about $3.6 billion in 2001 to about $6.0 billion in 2006, a compound annual growth rate of 10.6%. North American sales of DC/DC converters in networking and telecom equipment are expected to grow from $2.3 billion in 1999 to $4.6 billion in 2004, a yearly compounded growth rate of 15.3%. The first quarter of 1995 was the first time that shipments of 3.3 V microprocessor and memory integrated chips (ICs) exceeded the shipments of 5 V parts. Once this transition occurred in memories and microprocessors, the demand for low-voltage power conversion began to grow at an increasing rate. The primary application of DC/DC converters is in computer power supplies and communication equipment.

The computer power supply shift from 5 V to 3.3 V for digital IC has taken over 5 years to occur, from the first indication that voltages below 5 V would be needed until the realization of volume sales. Now that the 5 V barrier has been broken, the trend toward lower and lower voltages is accelerating. Within 2 years, 2.5 V parts are expected to become common with the introduction of the next-generation microprocessors. In fact, current plans for next-generation microprocessors call for a dual voltage of 1.5/2.5 V with 1.5 V used for the memory bus and 2.5 V used for logic functions. Within another few years, voltages are expected to move as low as 0.9 V, with mainstream, high-volume parts operating at 1.5 V.

A low-voltage plus high-current DC power supply is urgently required in the next-generation computer and communications equipment. The first idea is to use a forward converter, which can perform low-voltage plus high-current output voltage. A modified circuit with dynamic clamp circuit is shown in Figure 1.1. The two diodes D_1 and D_2 can be normal rectifier diode, rectifier Schottky diode, or MOSFET. Figure 1.2 shows the efficiency gain of the following three types of forward converters needed to construct a low-voltage high-current power supply:

- Forward converter using traditional diodes
- Forward converter using Schottky diode
- Synchronous rectifier using low forward-resistance MOSFET

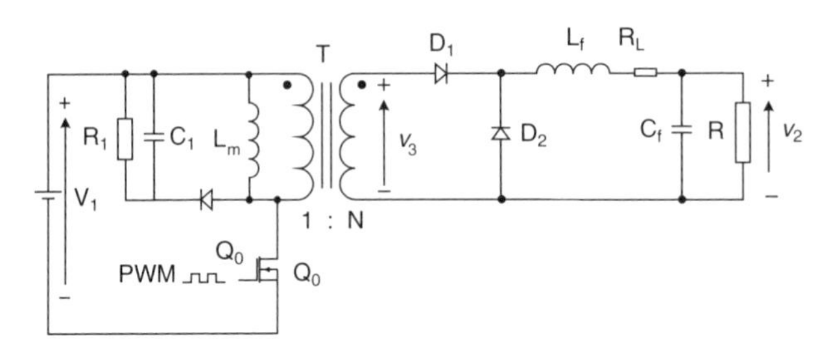

FIGURE 1.1
Forward converter with dynamic clamp circuit.

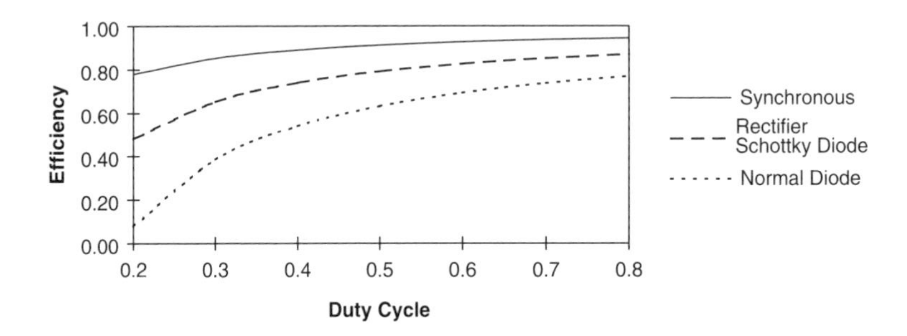

FIGURE 1.2
Efficiencies of different types of forward converter.

As the operating voltages ratchet downward, the design of rectifiers requires more attention because the devices forward-voltage drop constitutes an increasing fraction of the output voltage. The forward-voltage drop across a switch-mode rectifier is in series with the output voltage, so losses in this rectifier will almost entirely determine its efficiency. The synchronous rectifier circuit has been designed primarily to reduce this loss.

1.1 Introduction

A synchronous rectifier is an electronic switching circuit that improves power-conversion efficiency by placing a low-resistance conduction path across the diode rectifier in a switch-mode regulator. For a 3.3 V power supply, the traditional diode-rectifier loss is significant with very low efficiency (say less than 70%). For step-down regulators with a 3.3 V output and 12 V battery input voltage, a 0.4 V forward voltage of a Schottky diode represents a typical efficiency penalty of about 12%, aside from other loss mechanisms. The losses

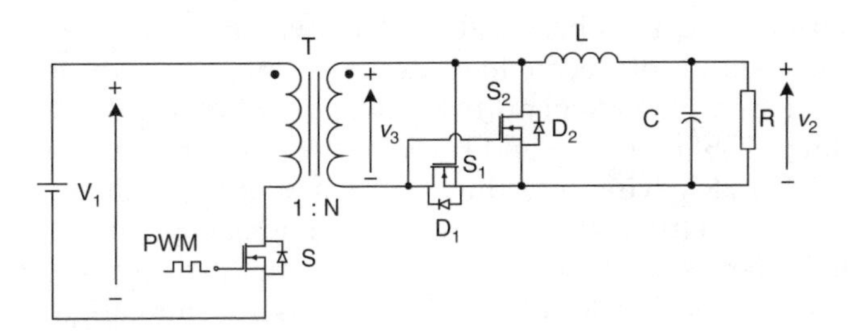

FIGURE 1.3
Synchronous rectifier converter with low-resistance MOSFET.

are not as bad at lower input voltages because the rectifier has a lower duty cycle and, thus, a shorter conduction time. However, the Schottky rectifier's forward drop is usually the dominant loss mechanism.

For an input voltage of 7.2 V and output of 3.3 V, a synchronous rectifier improves on the Schottky diode rectifier's efficiency by around 4%. Figure 1.1 also shows that, as output voltage decreases, the synchronous rectifier provides even larger gains in efficiency.

A practical circuit arrangement of a synchronous rectifier (SR) DC/DC converter with purely resistive load is shown in Figure 1.3. It has one MOSFET switch S on the primary side of the transformer. Two MOSFETs S_1 and S_2 on the secondary side of the transformer are functioning as the synchronous rectifier. T is the isolating transformer with a turn ratio of 1:N. An L-C circuit is the low-pass filter and R is the load. V_1 is the input voltage and V_2 is the output voltage. The main switch S is driven by a PWM pulse-train signal. Repeating frequency f and turn-on duty cycle k of the PWM signal can be adjusted.

When the PWM signal is in the positive state, the main switch S conducts. The primary voltage of the transformer is V_1, and subsequently the secondary voltage of the transformer is $v_3 = NV_1$. In the mean time the MOSFET S_1 is forward biased, so it turns ON and inversely conducts.

When the PWM signal is in the negative state, the main switch S is switched off. The voltage of the transformer, v_3, at this moment in time is approximately $-Nv_{C1}$. At the mean time the MOSFET S_2 is forward biased, so it turns ON and inversely conducts. It functions as free-wheeling diode and lets the load current remain continuous through the filter L–C and load R.

A lot of papers in literature with practical hardware circuit achievements on synchronous rectifier have been presented about the recent *IEEE Transactions* and *IEE Proceedings*. The paper "Evaluation of Synchronous-Rectification Efficiency Improvement Limits in Forward Converters" supported by Virginia Power Electronics Center is one of the few outstanding research publications on synchronous rectifiers (Jovanovic et al., 1995). This chapter provides a practical design of a 3.3V/20A FSR (forward synchronous rectifier) with an efficiency of 85.5%. Similarly, another paper in 1993 also showed the principle of a RCD clamp forward converter with an efficiency of 87.3%

at low output current (Cobos et al., 1993). Two Japanese researchers from Kumamoto Institute of Technology designed an FSR with an additional winding and switching element that is able to hold the gate charge for the freewheeling MOSFET (Sakai and Harada, 1995). Their experimental results for a 5 V/10 A SR gave a maximum efficiency of approximately 91% at a load of 7 A and an efficiency of 89% at 10 A. Another comparable FSR project was made by James Blanc from Siliconix Incorporated (Blanc, 1991). In his paper, he has included a lot of practical and useful simulation and experimental waveform data from his 3.3 V/10 A FSR. As the output voltage decreased, the operating efficiency decreased. Until now, no recent paper has been published on any practical hardware FSR that is able to provide 1.8 V/20 A output current at high efficiency.

Analysis and design of DC/DC converters has been the subject of many papers in the past. From the moment averaging techniques were used to model these converters, interest has been focused on finding the best approach to analyze and predict the behavior of the averaged small signal or large signal models. The main difficulty encountered is that the converter models are multiple-input multiple-output nonlinear systems and thus, using the well-known transfer function control design approach is not straightforward. The most common approach has been that of considering the linearized small signal model of these converters as a multi-loop system, with an outer voltage loop and an inner current loop. Since the current loop has a much faster response than that of the outer loop, the analysis is greatly simplified and the transfer functions obtained allow the designer to predict the closed loop behavior of the system. Another approach in analysis and design has been that of state-space techniques where the linearized state-space equations are used together with design technique such as pole placement or optimal control.

Synchronous rectifier DC/DC converters are called the fifth generation converters. The developments in microelectronics and computer science require power supplies with low output voltage and strong current. Traditional diode bridge rectifiers are not available for this requirement. Soft-switching technique can be applied in synchronous rectifier DC/DC converters. We have created converters with very low voltage (5 V, 3.3 V, and 1.8 ~ 1.5 V) and strong current (30 A, 60 A, 200 A) and high power transfer efficiency (86%, 90%, 93%). In this section new circuits different from the ordinary synchronous rectifier DC/DC converters are introduced:

- Flat transformer synchronous rectifier Luo-converter
- Active clamped flat transformer synchronous rectifier Luo-converter
- Double current synchronous rectifier Luo-converter with active clamp circuit
- Zero-current-switching synchronous rectifier Luo-converter
- Zero-voltage-switching synchronous rectifier Luo-converter

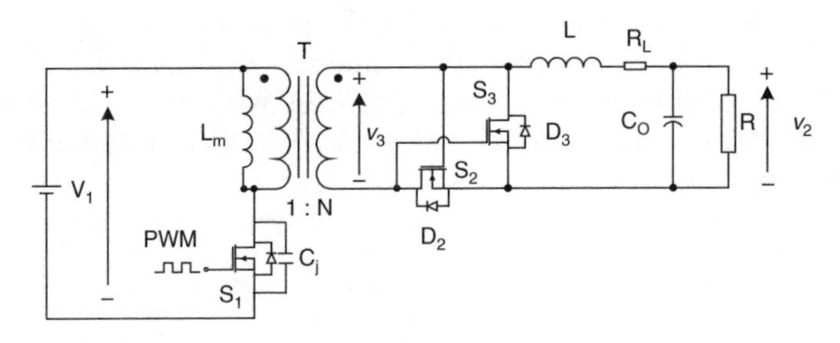

FIGURE 1.4
Flat transformer synchronous rectifier Luo-converter.

1.2 Flat Transformer Synchronous Rectifier Luo-Converter

The flat transformer is a new design for AC/AC energy conversion. Since its structure is very compatible and well-shielded, its size is very small and likely a flan cardboard. Applying frequency can be 100 KHz to 5 MHz, its power density can be as high as 300 W/inch^3. Therefore, it is a good component to use to construct the synchronous rectifier DC/DC converter.

The flat transformer SR DC/DC Luo-converter is shown in Figure 1.4. The switches S_2 and S_3 are the low-resistance MOSFET devices with resistance R_S (6 to 8 mΩ). Since we use a flat transformer, its leakage inductance L_m, L_m = 1 nH, and resistance are small. Other parameters are C_j = 50 ~ 100 nF, R_L = 2 mΩ, L = 5 μH, C_O = 10 μF. The input voltage is V_1 = 30 VDC and output voltage is V_2, the output current is I_O. The transformer turn ratio is N that is usually much smaller than unity in SR DC/DC converters, e.g., N = 1:12 or 1/12. The repeating period is $T = 1/f$ and conduction duty is k. There are four working modes:

- Transformer is in magnetizing
- Forward on
- Transformer is in demagnetizing
- Switched off

1.2.1 Transformer Is in Magnetizing Process

The natural resonant frequency is

$$\omega = \frac{1}{\sqrt{L_m C_j}} \tag{1.1}$$

where the L_m is the leakage inductance of the primary winding, the C_j is the drain-source junction capacitance of the main switch MOSFET S.

If C_j is very small in nF, its charging process is very quickly completed. The primary current increases with slope V_1/L_m, then the time interval of this period can be estimated

$$t_1 = \frac{L_m}{V_1} NI_O \tag{1.2}$$

This is the process used to establish the primary current from 0 to rated value NI_O.

1.2.2 Switching-On

Switching-on period is controlled by the PWM signal, therefore,

$$t_2 \approx kT \tag{1.3}$$

1.2.3 Transformer Is in Demagnetizing Process

The transformer demagnetizing process is estimated in

$$t_3 = \sqrt{L_m C_j}\left[\frac{\pi}{2} + \frac{V_1}{\sqrt{V_1^2 + \frac{L_m}{C_j}(NI_O)^2}}\right] \tag{1.4}$$

When the main switch is switching-off there is a voltage stress, which can be very high. The voltage stress is dependent on the energy stored in the inductor and the capacitor:

$$V_{peak} = \sqrt{\frac{L_m}{C_j}} NI_O \tag{1.5}$$

The voltage stress peak value can be tens to hundreds of volts since C_j is small.

1.2.4 Switching-Off

The switch-off period is controlled by the PWM signal, therefore,

$$t_4 \approx (1-k)T \tag{1.6}$$

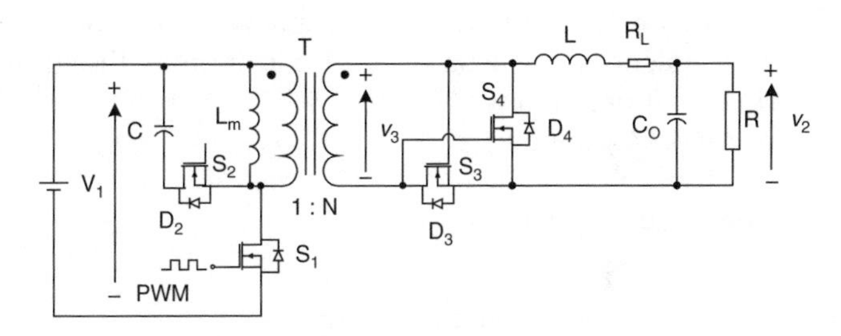

FIGURE 1.5
Active clamped flat transformer synchronous rectifier Luo-converter.

1.2.5 Summary

Average output voltage V_2 and input current I_1 are

$$V_2 = kNV_1 - (R_L + R_S + \frac{L_m}{T} N^2) I_O \tag{1.7}$$

and

$$I_1 = kNI_O \tag{1.8}$$

The power transfer efficiency:

$$\eta = \frac{V_2 I_O}{V_1 I_1} = 1 - \frac{R_L + R_S + \frac{L_m}{T} N^2}{kNV_1} I_O \tag{1.9}$$

When we set V_1 = 30 V, N = 1/12, k = 0.72–0.85, the frequency f = 150 to 200 kHz, we obtained the V_2 = 1.8 V, I_O = 0 to 30 A, Volume = 2.5 (in.^{3}). The average power transfer efficiency is 92.3% and the maximum PD is 21.6 W/in.^{3}.

1.3 Active Clamped Synchronous Rectifier Luo-Converter

Active clamped flat transformer SR Luo-converter is shown in Figure 1.5. The clamped circuit effectively suppresses the voltage stress during the main switch turn-off.

Comparing the circuit in Figure 1.5 with the circuit in Figure 1.4, one more switch S_2 is set in the primary side. It is switched-on and -off exclusively to the main switch S_1. When S_1 is turning-off, S_2 is switching-on. A large clamp capacitor C is connected in the primary winding to absorb the energy stored in the leakage inductor L_m. Since the clamp capacitor C is much larger than the drain-source capacitor C_j by usually hundreds of times, the stress voltage peak value remains at only a few volts.

There are four working modes:

- Transformer is in magnetizing
- Forward on
- Transformer is in demagnetizing
- Switched off

1.3.1 Transformer Is in Magnetizing

The natural resonant frequency is

$$\omega = \frac{1}{\sqrt{L_m C_j}} \tag{1.10}$$

where the L_m is the leakage inductance of the primary winding, the C_j is the drain-source junction capacitance of the main switch MOSFET S.

If C_j is very small in nF, its charging process is very quickly completed. The primary current increases with slope V_1/L_m, then the time interval of this period can be estimated

$$t_1 = \frac{L_m}{V_1} NI_O \tag{1.11}$$

This is the process to establish the primary current from 0 to rated value NI_O.

1.3.2 Switching-On

Switching-on period is controlled by the PWM signal, therefore,

$$t_2 \approx kT \tag{1.12}$$

1.3.3 Transformer Is in Demagnetizing

The transformer demagnetizing process is estimated by

$$t_3 = \sqrt{L_m C}[\frac{\pi}{2} + \frac{V_1}{\sqrt{V_1^2 + \frac{L_m}{C}(NI_O)^2}}] \tag{1.13}$$

where C is the active clamp capacitor in μF. The voltage stress depends on the energy stored in the inductor and the capacitor:

$$V_{peak} = \sqrt{\frac{L_m}{C}} N I_O \tag{1.14}$$

The voltage stress peak value is very small since capacitor C is large, measured in μF.

1.3.4 Switching-Off

Switching-off period is controlled by the PWM signal, therefore,

$$t_4 \approx (1-k)T \tag{1.15}$$

1.3.5 Summary

Average output voltage is V_2 and input current is I_1:

$$V_2 = kNV_1 - (R_L + R_S + \frac{L_m}{T} N^2) I_O \tag{1.16}$$

and

$$I_1 = kNI_O \tag{1.17}$$

The power transfer efficiency:

$$\eta = \frac{V_2 I_O}{V_1 I_1} = 1 - \frac{R_L + R_S + \frac{L_m}{T} N^2}{kNV_1} I_O \tag{1.18}$$

When we set V_1 = 30 V, N = 1/12, the frequency f = 150 to 200 kHz, k = 0.72–0.85, we obtained the V_2 = 1.8 V, I_O = 0 to 30 A, volume = 2.5 (in.^{3}). The average power transfer efficiency is 92.3% and the maximum power density (PD) is 21.6 W/in.^{3}.

1.4 Double Current Synchronous Rectifier Luo-Converter

The converter in Figure 1.5 likes a half wave rectifier. The double current synchronous rectifier Luo-converter with active clamp circuit is shown in

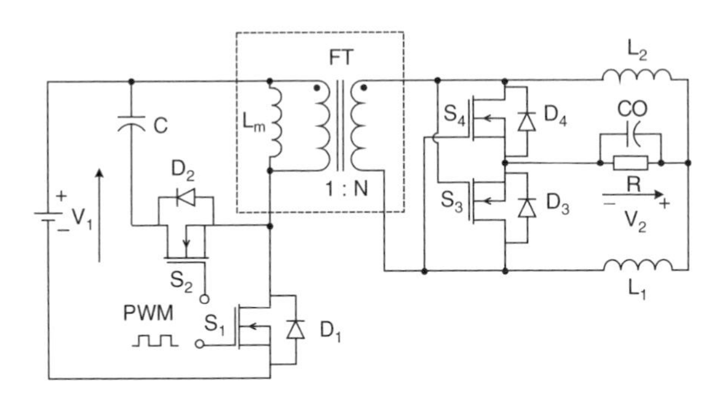

FIGURE 1.6
Double-current synchronous rectifier Luo-converter.

Figure 1.6. The switches S_3 and S_4 are the low-resistance MOSFET devices with very low resistance R_S (6 to 8 mΩ). Since S_3 and S_4 plus L_1 and L_2 form a double current circuit and S_2 plus C is the active clamp circuit, this converter likes a full wave rectifier and obtains strong output current. Other parameters are $C = 1$ μF, $L_m = 1$ nH, $R_L = 2$ mΩ, $L = 5$ μH, $C_O = 10$ μF. The input voltage is $V_1 = 30$ VDC and output voltage is V_2, the output current is I_O. The transformer turn ratio is $N = 1{:}12$. The repeating period is $T = 1/f$ and conduction duty is k. There are four working modes:

- Transformer is in magnetizing
- Forward on
- Transformer is in demagnetizing
- Switched off

1.4.1 Transformer Is in Magnetizing

The natural resonant frequency is

$$\omega = \frac{1}{\sqrt{L_m C_j}} \tag{1.19}$$

where the L_m is the leakage inductance of the primary winding, the C_j is the drain-source junction capacitance of the main switch MOSFET S.

If C_j is very small in nF, its charging process is very quickly completed. The primary current increases with slope V_1/L_m, and the time interval of this period can be estimated

$$t_1 = \frac{L_m}{V_1} N I_O \tag{1.20}$$

This is the process used to establish the primary current from 0 to rated value NI_O.

1.4.2 Switching-On

Switching-on period is controlled by the PWM signal, therefore,

$$t_2 \approx kT \tag{1.21}$$

1.4.3 Transformer Is in Demagnetizing

The transformer demagnetizing process is estimated in

$$t_3 = \sqrt{L_m C}\left[\frac{\pi}{2} + \frac{V_1}{\sqrt{V_1^2 + \frac{L_m}{C}(NI_O)^2}}\right] \tag{1.22}$$

When the main switch is switching-off there is a very low voltage stress since the active clamp circuit is applied.

1.4.4 Switching-Off

Switching-off period is controlled by the PWM signal, therefore,

$$t_4 \approx (1-k)T \tag{1.23}$$

1.4.5 Summary

Average output voltage V_2 and input current I_1 are

$$V_2 = kNV_1 - (R_L + R_S + \frac{L_m}{T}N^2)I_O \tag{1.24}$$

and

$$I_1 = kNI_O \tag{1.25}$$

The power transfer efficiency is

$$\eta = \frac{V_2 I_O}{V_1 I_1} = 1 - \frac{R_L + R_S + \frac{L_m}{T}N^2}{kNV_1} I_O \tag{1.26}$$

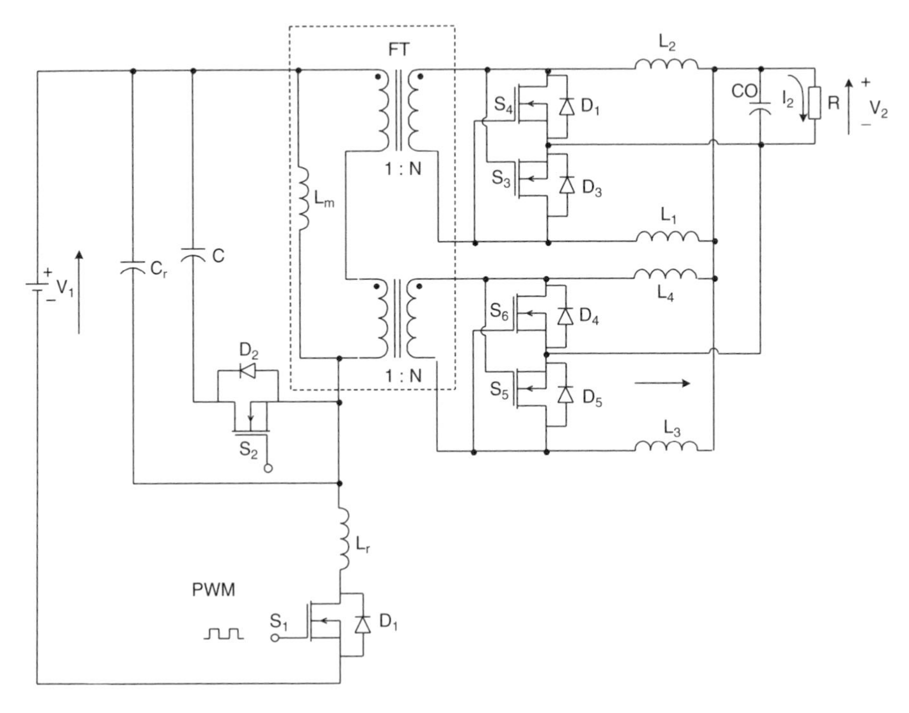

FIGURE 1.7
ZCS synchronous rectifier Luo-converter.

When we set the frequency f = 200 to 250 kHz, we obtained the V_2 = 1.8 V, N = 12, I_O = 0 to 35 A, volume = 2.5 (in.^{3}). The average power transfer efficiency is 94% and the maximum PD is 25 W/in^3.

1.5 Zero-Current-Switching Synchronous Rectifier Luo-Converter

The zero-current-switching (ZCS) synchronous rectifier Luo-converter is shown in Figure 1.7. Since the power loss across the main switch S_1 is high in the double current SR Luo-converter, we designed the ZCS, DC SR Luo-Converter shown in Figure 1.7. This converter is based on the DC SR Luo-converter plus ZCS technique. It employs a double core flat transformer. There are four working modes:

- Transformer is in magnetizing
- Resonant period
- Transformer is in demagnetizing
- Switched off

1.5.1 Transformer Is in Magnetizing

The ZCS resonant frequency is

$$\omega_r = \frac{1}{\sqrt{L_r C_r}} \tag{1.27}$$

The normalized impedance is

$$Z_r = \sqrt{\frac{L_r}{C_r}} \tag{1.28}$$

The shift-angular distance is

$$\alpha = \sin^{-1}\left(\frac{I_1 Z_r}{V_1}\right) \tag{1.29}$$

where the L_r is the resonant inductor and the C_r is the resonant capacitor.

The primary current increases with slope V_1/L_r, then the time interval of this period can be estimated

$$t_1 = \frac{I_1 L_r}{V_1} \tag{1.30}$$

1.5.2 Resonant Period

The resonant period is,

$$t_2 = \frac{1}{\omega_r}(\pi + \alpha) \tag{1.31}$$

1.5.3 Transformer Is in Demagnetizing

The transformer demagnetizing process is estimated in

$$t_3 = \frac{V_1(1 + \cos\alpha)C_r}{I_1} \tag{1.32}$$

When the main switch is switching-off there is a very low voltage stress since active clamp circuit is applied.

1.5.4 Switching-Off

Switching-off period is controlled by the PWM signal, therefore,

$$t_4 = \frac{V_1(t_1 + t_2)}{V_2 I_1}(I_L + \frac{V_1}{Z_r}\frac{\cos\alpha}{\pi/2+\alpha}) - (t_1 + t_2 + t_3) \tag{1.33}$$

1.5.5 Summary

Average output voltage V_2 and input current I_1 are

$$V_2 = kNV_1 - (R_L + R_S + \frac{L_r + L_m}{T}N^2)I_O \tag{1.34}$$

and

$$I_1 = kNI_O \tag{1.35}$$

The power transfer efficiency is

$$\eta = \frac{V_2 I_O}{V_1 I_1} = 1 - \frac{R_L + R_S + \frac{L_r + L_m}{T}N^2}{kNV_1} I_O \tag{1.36}$$

Since L_r is larger than L_m, therefore L_m can be ignored in the above formulae.

When we set the $V_1 = 60$ V and frequency $f = 200$ to 250 kHz, we obtained the $V_2 = 1.8$ V, $N = 1/12$, $I_O = 0$ to 60 A, volume = 4 (in.^{3}). The average power transfer efficiency is 94.5% and the maximum PD is 27 W/in.^{3}.

1.6 Zero-Voltage-Switching Synchronous Rectifier Luo-Converter

The zero-voltage-switching synchronous rectifier Luo-converter is shown in Figure 1.8, which is derived from the DC SR Luo-converter plus ZVS technique. It employs a double core flat transformer. There are four working modes:

- Transformer is in magnetizing
- Resonant period
- Transformer is in demagnetizing
- Switched off

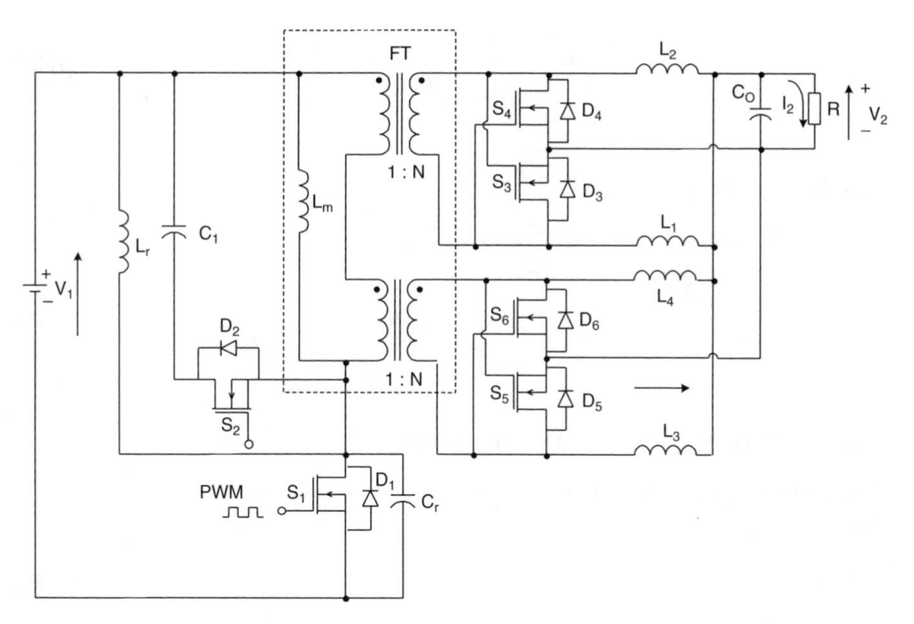

FIGURE 1.8
ZVS synchronous rectifier Luo-converter.

1.6.1 Transformer Is in Magnetizing

The ZVS resonant frequency is

$$\omega_r = \frac{1}{\sqrt{L_r C_r}} \tag{1.37}$$

The normalized impedance is

$$Z_r = \sqrt{\frac{L_r}{C_r}} \tag{1.38}$$

The shift-angular distance is

$$\alpha = \sin^{-1}\left(\frac{V_1}{Z_r I_1}\right) \tag{1.39}$$

where the L_r is the resonant inductor and the C_r is the resonant capacitor.

The switch voltage increases with slope I_1/C_r, then the time interval of this period can be estimated

$$t_1 = \frac{V_1 C_r}{I_1} \tag{1.40}$$

1.6.2 Resonant Period

The resonant period is,

$$t_2 = \frac{1}{\omega_r}(\pi + \alpha) \tag{1.41}$$

1.6.3 Transformer Is in Demagnetizing

The transformer demagnetizing process is estimated in

$$t_3 = \frac{I_1(1 + \cos\alpha)L_r}{V_1} \tag{1.42}$$

While the main switch is switching-off there is a very low voltage stress since active clamp circuit is applied.

1.6.4 Switching-Off

Switching-off period is controlled by the PWM signal, therefore,

$$t_4 = \frac{t_1 + t_2 + t_3}{\frac{V_1}{V_2} - 1} \tag{1.43}$$

1.6.5 Summary

Average output voltage V_2 and input current I_1 are

$$V_2 = kNV_1 - (R_L + R_S + \frac{L_r // L_m}{T} N^2) I_O \tag{1.44}$$

and

$$I_1 = kNI_O \tag{1.45}$$

The power transfer efficiency is

$$\eta = \frac{V_2 I_O}{V_1 I_1} = 1 - \frac{R_L + R_S + \frac{L_r // L_m}{T} N^2}{kNV_1} I_O \qquad (1.46)$$

Since L_r is larger than L_m, therefore L_r can be ignored in the above formulae.

When we set the V_1 = 60 V and frequency f = 200 to 250 kHz, we obtained the V_2 = 1.8 V, N = 12, I_O = 0 to 60 A, volume = 4 (in.^{3}). The average power transfer efficiency is 94.5% and the maximum PD is 27 W/in.^{3}.

Bibliography

Blanc, J., Practical Application of MOSFET Synchronous Rectifiers, in *Proceedings of the Telecom-Eng. Conference INTELEC'1991*, Japan, 1991, p. 495.

Chakrabarty, K., Poddar, G., and Banerjee, S., Bifurcation behavior of the buck converter, *IEEE Trans. Power Electronics*, 11, 439, 1996.

Cobos, J.A., O. Grcia, J., and Sebastian, J.U., RCD clamp PWM forward converter with self driven synchronous rectification, in *Proceedings of IEEE International Telecommunication Conference INTELEC'93*, 1993, p. 1336.

Cobos, J.A., Sebastian, J., Uceda, J., Cruz, E. de la, and Gras, J.M., Study of the applicability of self-driven synchronous rectification to resonant topologies, in *Proceedings of IEEE Power Electronics Specialists Conference*, 1992, p. 933.

Garofalo, F., Marino, P., Scala, S., and Vasca, F., Control of DC-DC converters with linear optimal feedback and nonlinear feedforward, *IEEE Trans. Power Electronics*, 9, 607, 1994.

Jovanovic, M.M., Zhang, M.T., and Lee, F.C., Evaluation of synchronous-rectification efficiency improvement limits in forward converters, *IEEE Transactions on Industrial Electronics*, 42, 387, 1995.

Leu, C.S., Hua, G., and Lee, F.C., Analysis and design of RCD clamp forward converter, in *Proceedings of the VPEC'92*, 1992, p. 198.

Luo, F.L. and Chua, L.M., Fuzzy logic control for synchronous rectifier DC/DC converter, in *Proceedings of the IASTED International Conference-ASC'00*, Canada, 2000, p. 24.

Mattavelli, P., Rossetto, L., and Spiazzi, G., General-purpose sliding-mode controller for DC/DC converter applications, *IEEE Trans. Power Electronics*, 8, 609, 1993.

Ogata, K., *Designing Linear Control Systems with MATLAB*, Prentice-Hall, Upper Saddle River, NJ, 1994.

Phillips, C.P. and Harbor, R., *Feedback Control System*, Prentice-Hall, 1991.

Rim, C.T., Joung, G.B., and Cho, G.H., Practical switch based state-space modeling of DC-DC converters with all parasitics, *IEEE Trans. Power Electronics*, 6, 611, 1991.

Sakai, E. and Harada, K., Synchronous Rectifier for Low Voltage Switching Converter, in *Proceedings of the Telecom-Eng. Conference INTELEC'1995*, Japan, 1995, p. 471.

Yamashita, N., Murakami, N., and Yachi, Toshiaki., Conduction power loss in MOSFET synchronous rectifier with parallel-connected Schottky barrier diode, *IEEE Transaction on Power Electronics*, 13, 667, 1998.

2

Multiple Energy-Storage Element Resonant Power Converters

2.1 Introduction

Multiple energy-storage element resonant power converters (x-element RPC) are the sixth generation converters. As the transfer power becomes higher and higher, traditional methods are unable to deliver large amounts of power from the source to the final actuators with high efficiency. In order to reduce the power losses during the conversion process the sixth generation converters — multiple energy-storage elements resonant power converters (x-element RPC), were created. They can be classified into two main groups

- DC/DC resonant converters
- DC/AC resonant inverters

Both groups consist of multiple energy-storage elements: two, three, or four elements. These energy-storage elements are passive parts: inductors and capacitors. They can be connected in series or parallel in various methods. The circuits of the multiple energy-storage elements converters are

- Eight topologies of two-element RPC shown in Figure 2.1.
- Thirty-eight topologies of three-element RPC shown in Figure 2.2.
- Ninety-eight topologies of four-element ($2L$-$2C$) RPC shown in Figure 2.3.

If no restriction such as $2L$-$2C$ for four-element RPC, the number of the topologies of four-element RPC can be very large. How to investigate the large quantity of converters is a task of vital importance. This problem was outstanding in the last decade of last century. Unfortunately, it was not paid very much attention. This generation of converters were not well discussed, only a limited number of papers were published in the literature.

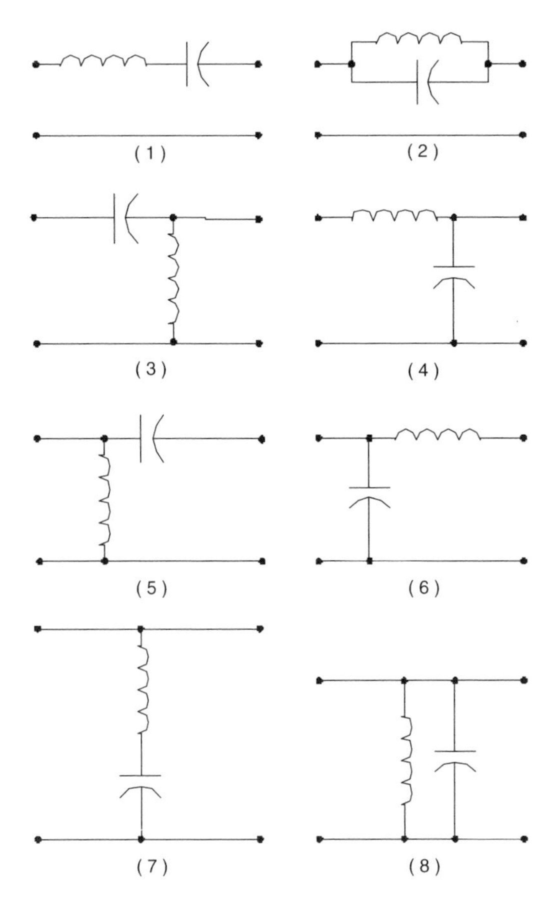

FIGURE 2.1
The eight topologies of two-element RPC.

2.1.1 Two-Element RPC

There are eight topologies of two-element RPCs shown in Figure 2.1. These topologies have simple circuit structure and minimal components. Consequently, they can transfer the power from source to end-users with higher power efficiency and lower power losses. A particular circuitry analysis will be carried out in the next section.

Usually, the two-element RPC has a very narrow response frequency band, which is defined as the frequency width Δf between the two half-power points.

Assume the resonant L-C circuit has inductance L and capacitance C, and the load is a pure resistive load R. The resonant frequency f_0 is

$$f_0 = \frac{\omega_0}{2\pi} = \frac{1}{2\pi\sqrt{LC}}$$

The normalized impedance Z_0 and quality factor Q is

$$Z_0 = \sqrt{\frac{L}{C}}$$

$$Q = \frac{Z_0}{R} = \sqrt{\frac{L}{C}} / R = \frac{\omega_0 L}{R} = \frac{1}{\omega_0 CR}$$

The frequency band width Δf is

$$\Delta f = \frac{f_0}{Q}$$

If the quality factor is large, the frequency band width is narrow.

The working point must be selected in the vicinity of the natural resonant radian frequency

$$\omega_0 = \frac{1}{\sqrt{LC}}$$

Another drawback is that the transferred waveform is usually not sinusoidal, i.e., the output waveform total harmonic distortion (THD) is not zero.

Since total power losses are mainly contributed by the power losses across the main switches using resonant conversion technique, the two-element RPC has a high power transferring efficiency.

2.1.2 Three-Element RPC

There are 38 topologies of three-element RPC that are shown in Figure 2.2. These topologies have one more component in comparison to the two-element RPC topologies. Consequently, they can transfer the power from source to end-users with higher power and high power transfer efficiency. A particular circuitry analysis will be carried out in the next chapter.

Usually, the three-element RPC has a much wider response frequency band, which is defined as the frequency width between the two half-power points. If the circuit is a low-pass filter, the frequency bands can cover the frequency range from 0 to the natural resonant radian frequency $\omega_0 = 1/\sqrt{LC}$. The working point can be selected in the much wider frequency width which is lower than the natural resonant radian frequency $\omega_0 = 1/\sqrt{LC}$. Another advantage over the two-element RPC topologies is that the transferred waveform can usually be sinusoidal, i.e., the output waveform THD is nearly zero. A well-known mono-frequency waveform transferring operation has very low EMI.

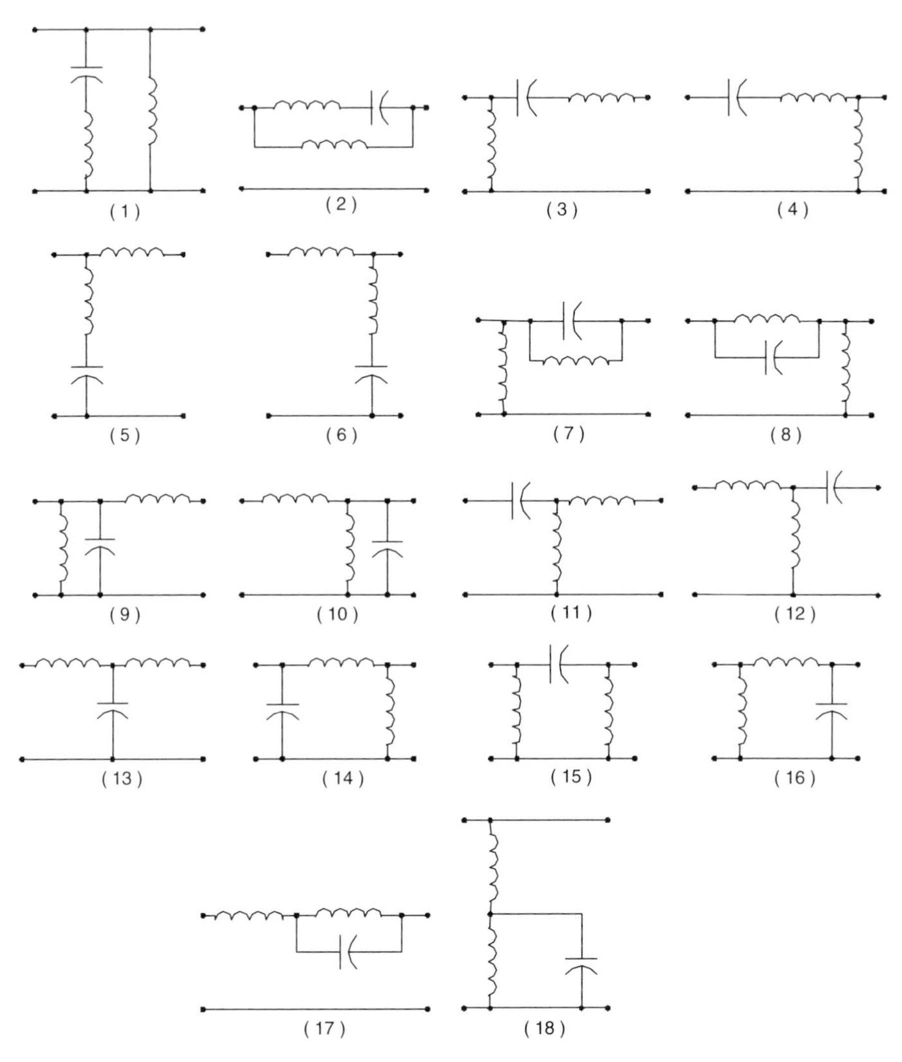

FIGURE 2.2
The thirty-eight topologies of three-element RPC.

2.1.3 Four-Element RPC

There are 98 topologies of four-element RPC (2L-2C) that are shown in Figure 2.3. Although these topologies have comparable complex circuit structures, they can still transfer the power from source to end-users with higher power efficiency and lower power losses. Particular circuitry analysis will be carried out in the next chapter.

Usually, the four-element RPC has a wide response frequency band, which is defined as the frequency width between the two half-power points. If the circuit is a low-pass filter, the frequency bands can cover the

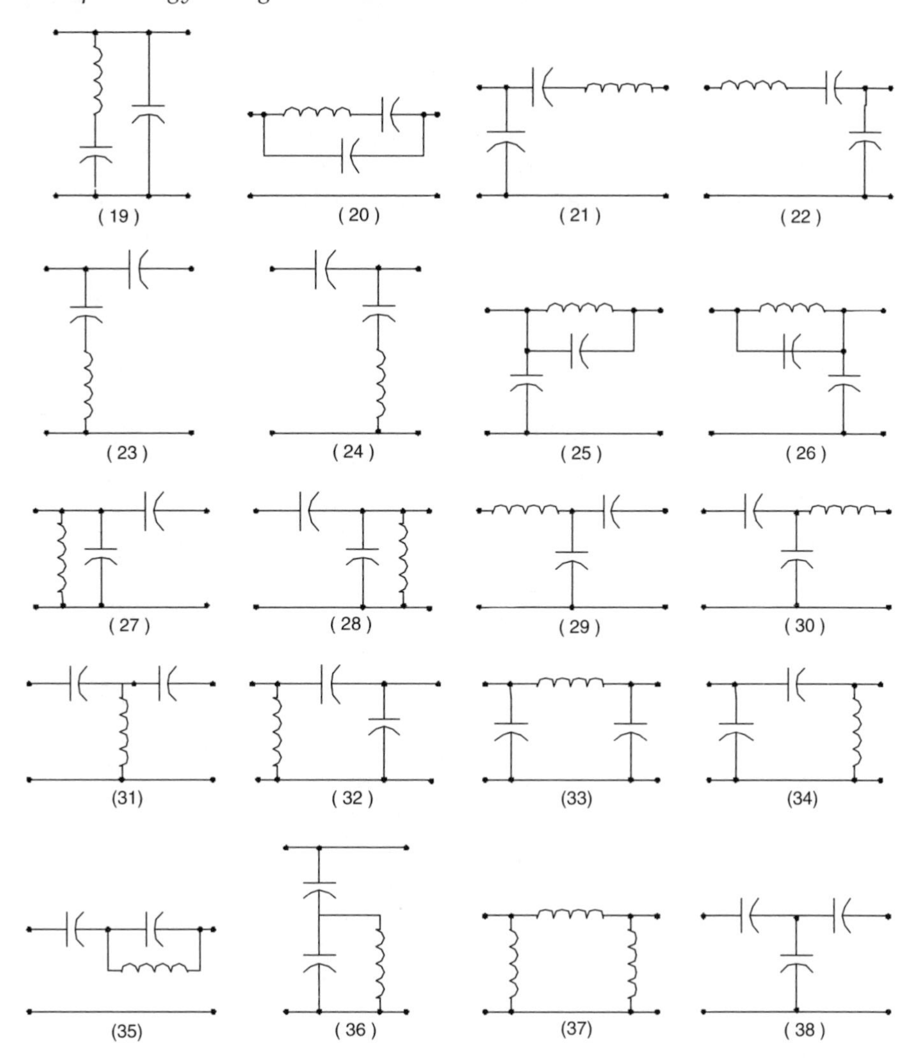

FIGURE 2.2 (continued)

frequency range from 0 to the high half-power point, which is definitely higher than the natural resonant radian frequency $\omega_0 = 1/\sqrt{LC}$. The working point can be selected in a wide area across (lower and higher than) the natural resonant radian frequency $\omega_0 = 1/\sqrt{LC}$. Another advantage is that the transferred waveform is sinusoidal, i.e., the output waveform THD is very close to zero. As is well-known, the mono-frequency-waveform transferring operation has a very low electromagnetic interference (EMI) and a reasonable electromagnetic susceptibility (EMS) and electromagnetic compatibility (EMC).

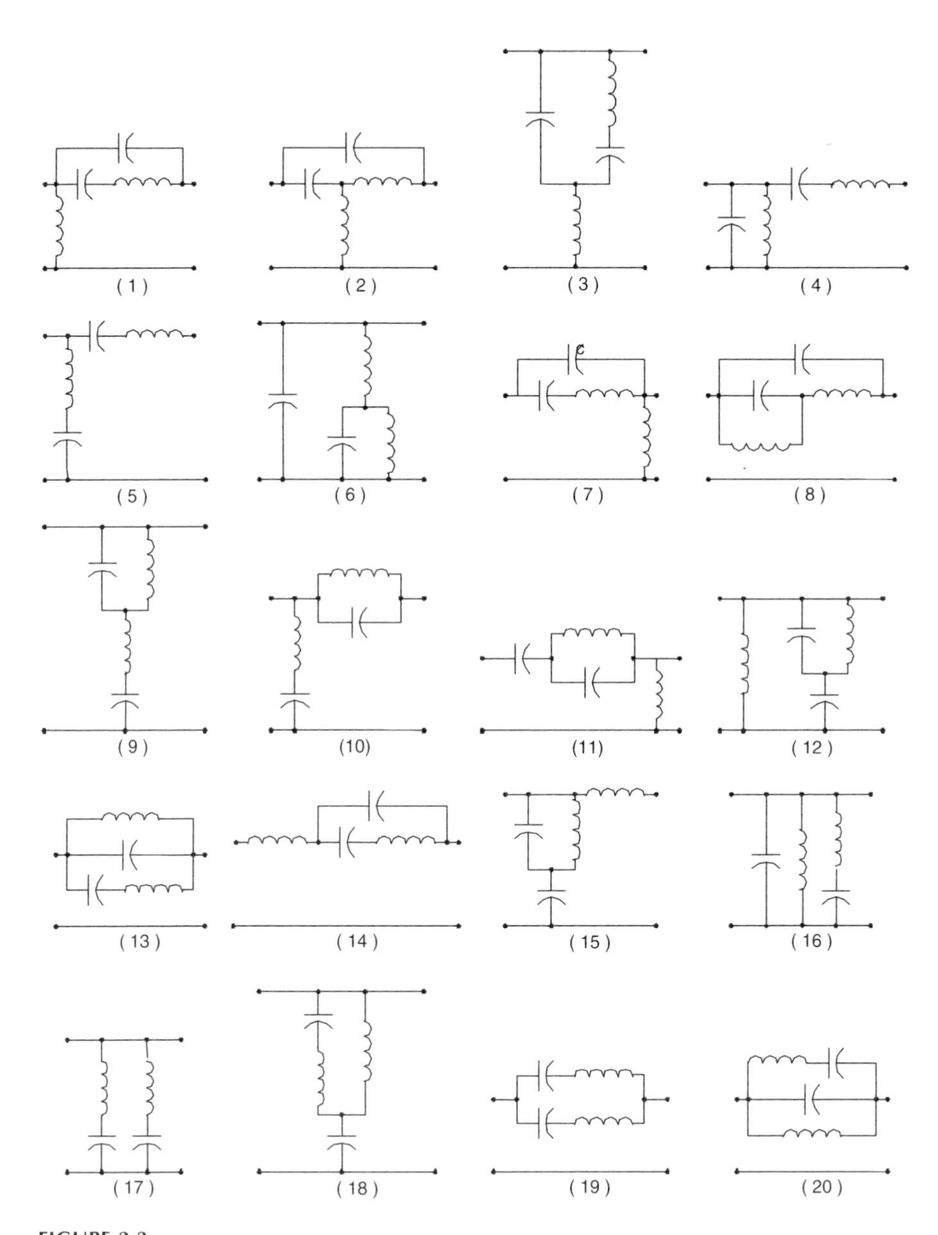

FIGURE 2.3
The ninety-eight topologies of four-element (2*L*-2*C*).

2.2 Bipolar Current and Voltage Source

Depending on the application, a resonant network can be low-pass filter, high-pass filter, or band-pass filter. For a large power transferring process,

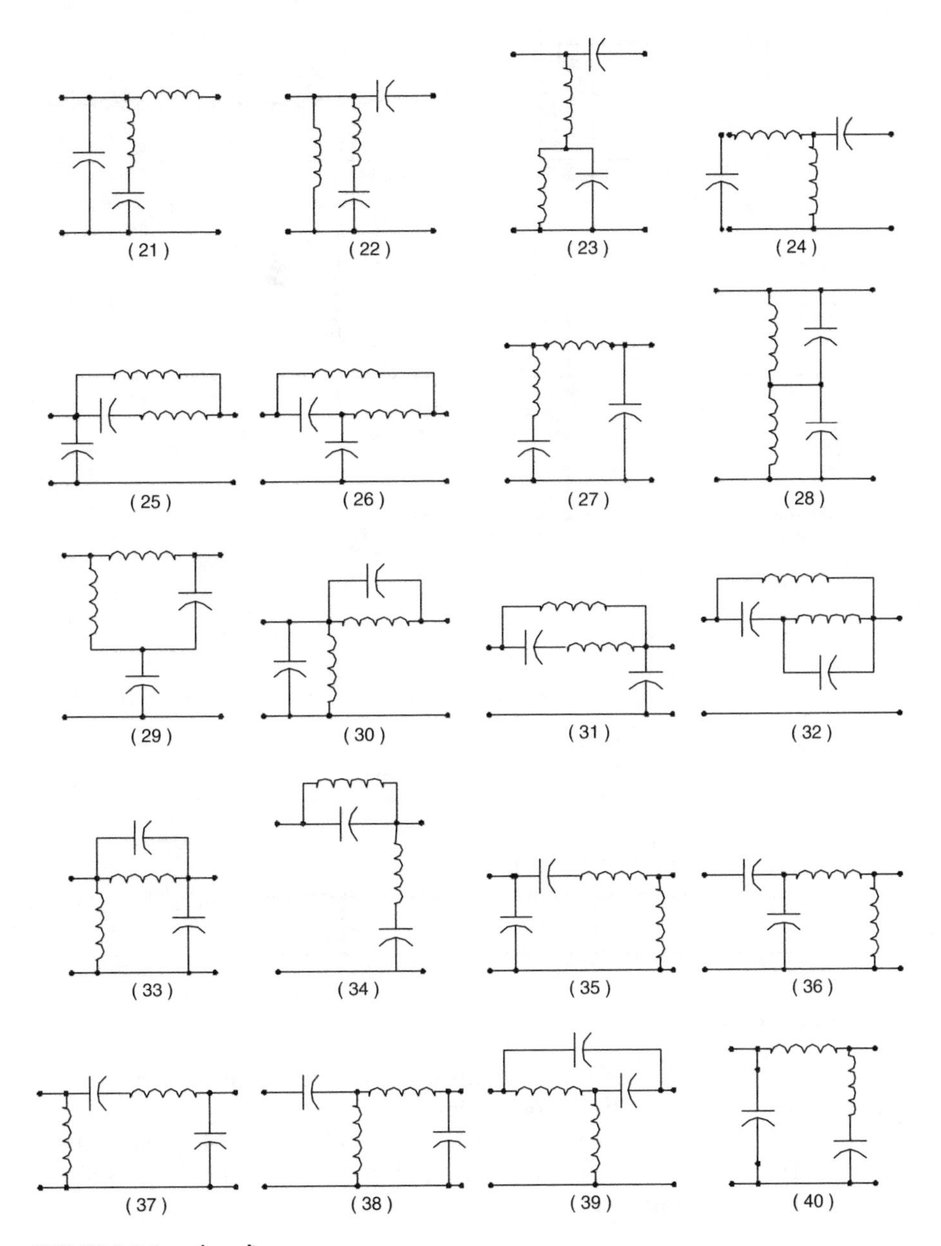

FIGURE 2.3 (continued)

a low-pass filter is usually employed. In this case, inductors are arranged in series arms and capacitors are arranged in shut arms. If the first component is an inductor, only voltage source can be applied since inductor current is continuous. Vice versa, if the first component is a capacitor, only current source can be applied since capacitor voltage is continuous. Unipolar current

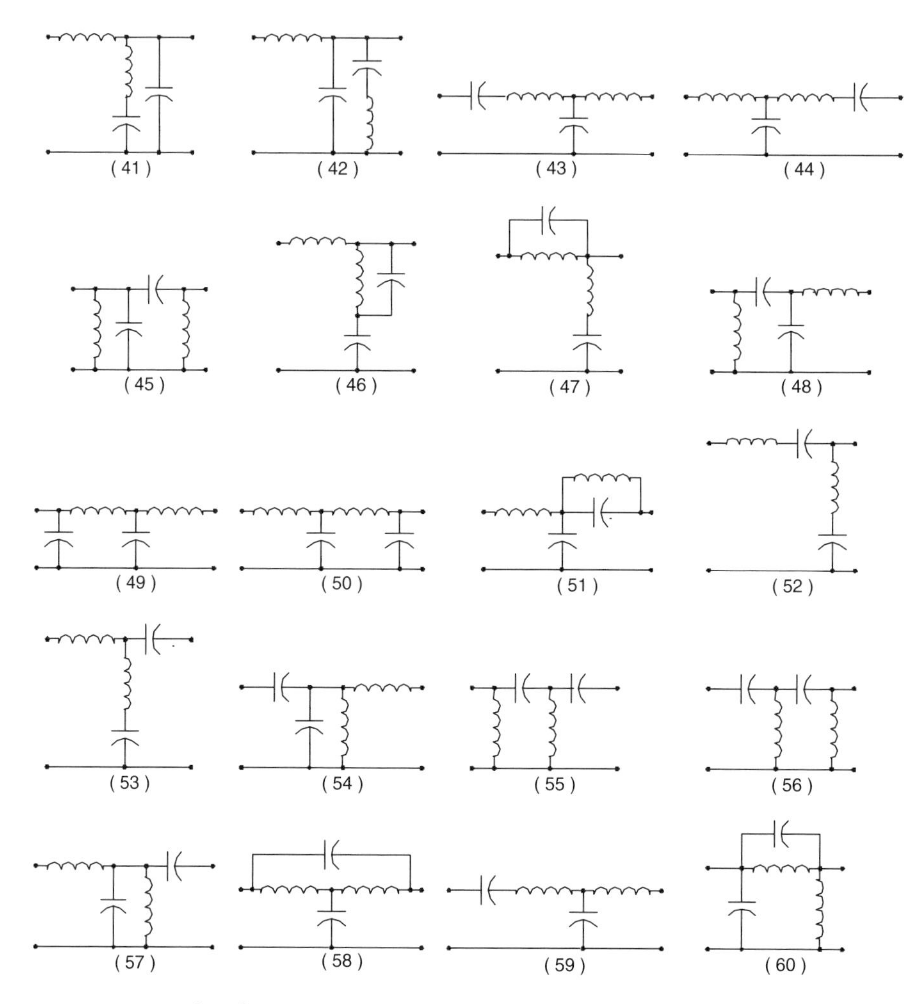

FIGURE 2.3 (continued)

and voltage source are easier to obtain using various pumps, such as buck pump, boost pump, and buck-boost pump.

Bipolar current and voltage sources are more difficult to obtain using these pumps. There are some fundamental bipolar current and voltage sources listed in the following sections.

2.2.1 Bipolar Voltage Source

There are various methods to obtain bipolar voltage sources using pumps.

2.2.1.1 Two Voltage Source Circuit

A bipolar voltage source using two voltage sources is shown in Figure 2.4. These two voltage sources have the same voltage amplitude and reverse

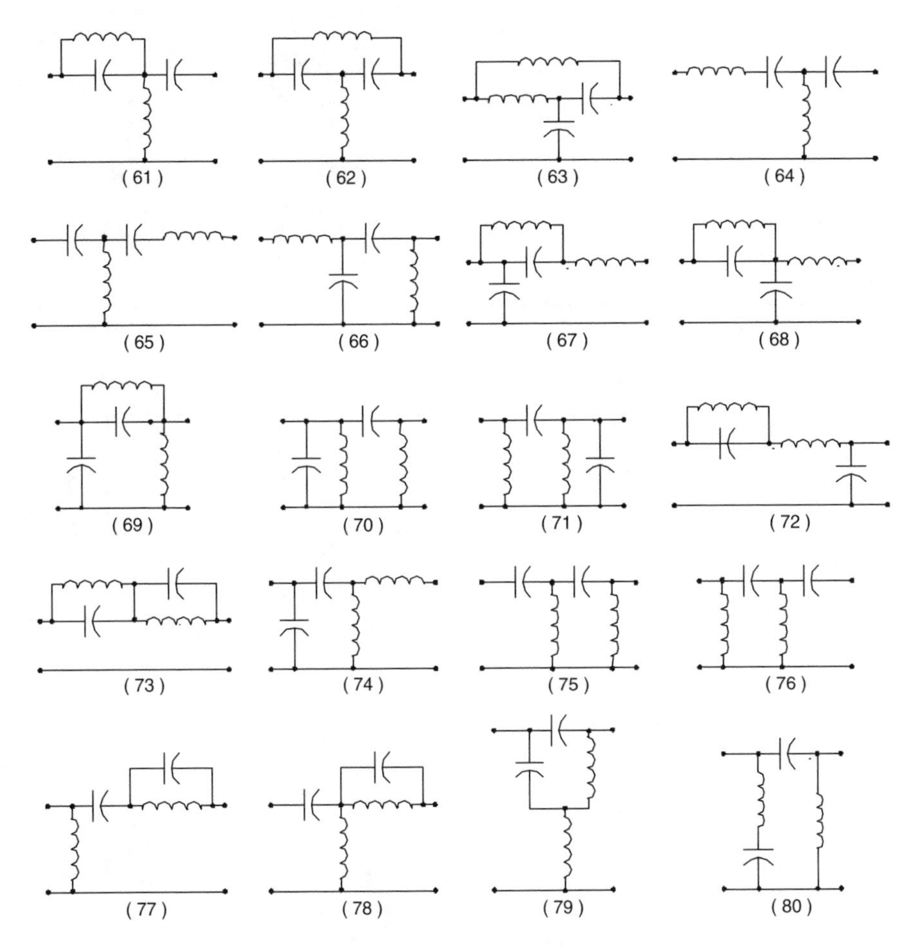

FIGURE 2.3 (continued)

polarity. There are two switches applied alternately switching-on or -off to supply positive and negative voltage to the network. In the figure, the load is a resistance R.

The circuit of this voltage source is likely a two-quadrant operational chopper. The conduction duty cycle for each switch is 50%. For safety reasons the particular circuitry design has to consider some small gap between the turn over (commutation) operations to avoid a short-circuit.

The repeating frequency is theoretically not restricted. For industrial applications, the operating frequency is usually arranged in the range between 10 kHz to 5 MHz depending on the application conditions.

2.2.1.2 *One Voltage Source Circuit*

A bipolar voltage source using single voltage source is shown in Figure 2.5. Since only one voltage source is applied, there are four switches applied

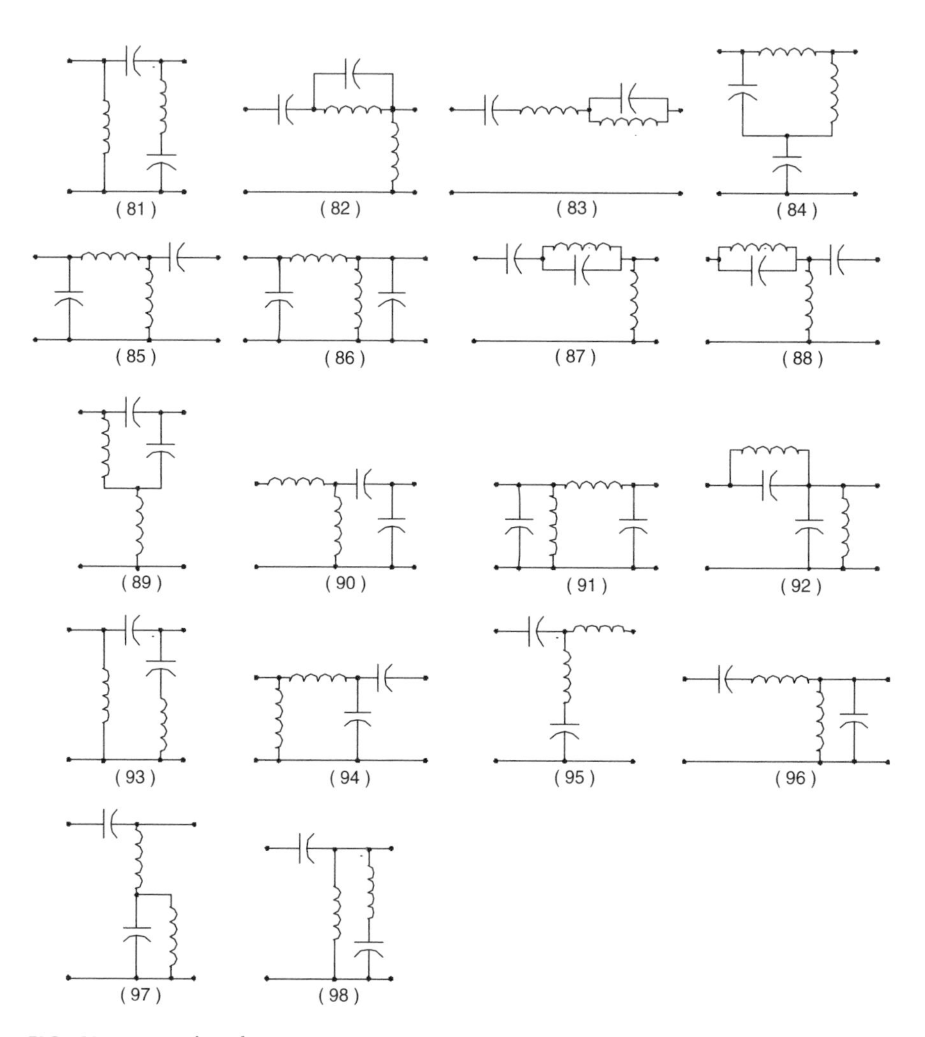

FIGURE 2.3 (continued)

alternately switching-on or switching-off to supply positive and negative voltage to the network. In the figure, the load is a resistance *R*.

The circuit of this voltage source is likely to be a four-quadrant operational chopper. The conduction duty cycle for each switch is 50%. For safety reasons the particular circuitry design has to consider some small gap between the turn over (commutation) operations to avoid a short-circuit.

The repeating frequency is theoretically not restricted. For industrial applications, the operating frequency is usually arranged in the range between 10 kHz to 5 MHz depending on the application conditions.

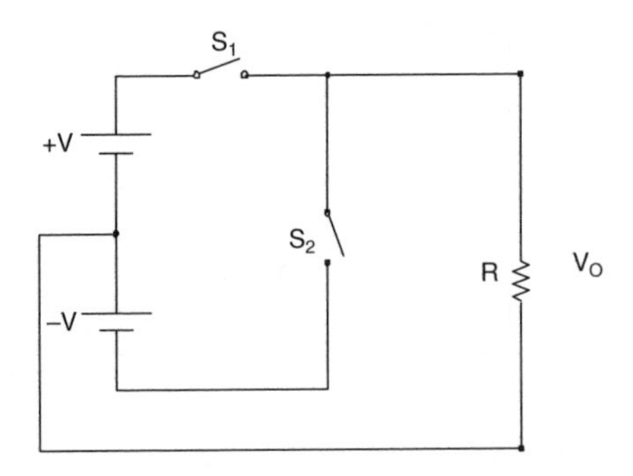

FIGURE 2.4
A bipolar voltage source using two voltage sources.

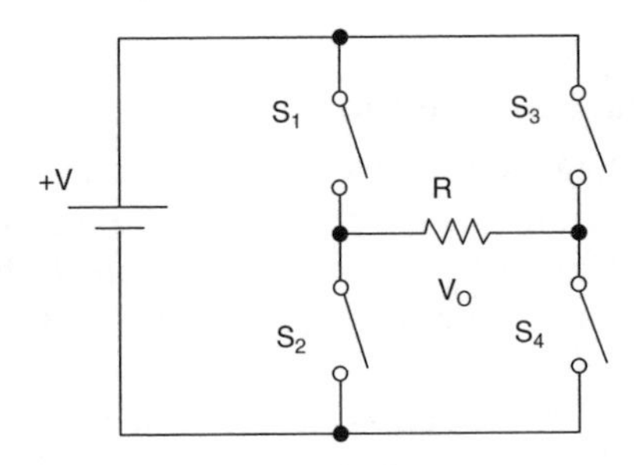

FIGURE 2.5
A bipolar voltage source using single voltage source.

2.2.2 Bipolar Current Source

There are various methods used to obtain bipolar current sources using the pumps.

2.2.2.1 Two Voltage Source Circuit

A bipolar current voltage source using two voltage sources is shown in Figure 2.6. These two voltage sources have the same voltage amplitude and reverse polarity. To obtain stable current each voltage source in series is connected by a large inductor. There are two switches applied alternately switching-on or switching-off to supply positive and negative current to the network. In Figure 2.6, the load is a resistance R.

The circuit of this current source is a two-quadrant operational chopper. The conduction duty cycle for each switch is 50%. For safety reasons the particular circuitry design has to consider some small gap between the turn over (commutation) operations to avoid a short-circuit.

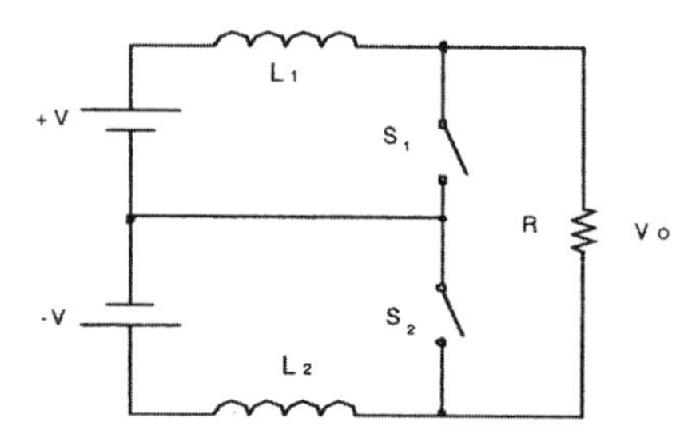

FIGURE 2.6
A bipolar current source using two voltage sources.

The repeating frequency is theoretically not restricted. For industrial applications, the operating frequency is usually arranged in the range between 10 kHz to 5 MHz depending on the application conditions.

2.2.2.2 One Voltage Source Circuit

A bipolar current voltage source using single voltage sources is shown in Figure 2.7. To obtain stable current the voltage source in series is connected by a large inductor. There are two switches applied alternately switching-on or -off to supply positive and negative current to the network. In the figure, the load is a resistance R.

The circuit of this current source is likely a two-quadrant operational chopper. The conduction duty cycle for each switch is 50%. For safety reasons the particular circuitry design has to consider some small gap between the turn over (commutation) operations to avoid a short-circuit.

The repeating frequency is theoretically not restricted. For industrial applications, the operating frequency is usually arranged in the range between 10 kHz to 5 MHz depending on the application conditions.

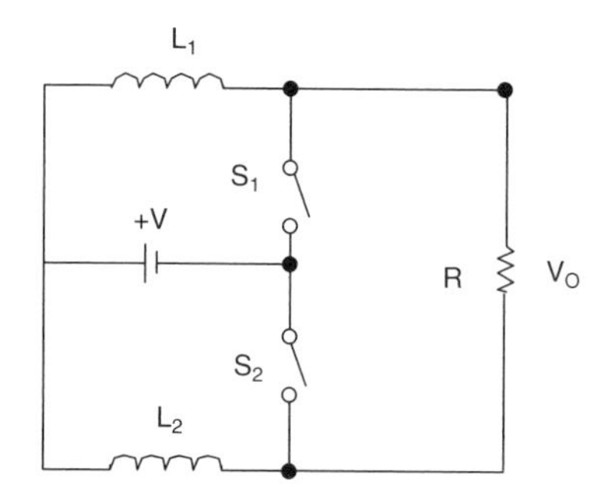

FIGURE 2.7
A bipolar current source using single voltage source.

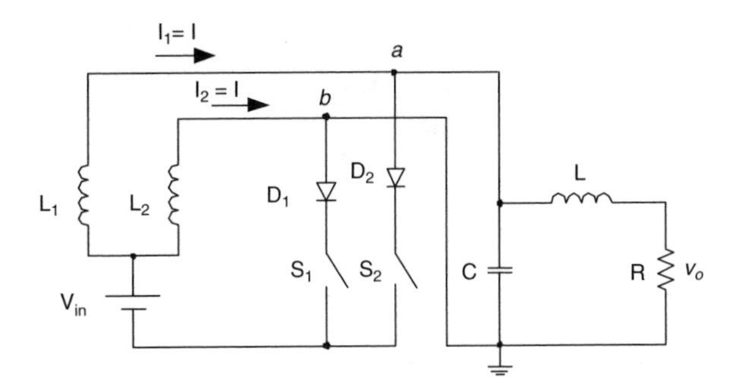

FIGURE 2.8
A two-element RPC.

2.3 A Two-Element RPC Analysis

This two-element RPC is the circuit number six in Figure 2.1. The network is a low-pass capacitor-inductor (CL) filter, and the first component is a capacitor. By previous analysis, the source should be a bipolar current source. Therefore, the circuit diagram of this two-element RPC is shown in Figure 2.8. To simplify the analysis, the load can be considered resistive R. The energy source V_{in} is chopped by two main switches S_1 and S_2, it is a bipolar current source applied to the two-element filter and load. The whole RPC equivalent circuit diagram is shown in Figure 2.9.

2.3.1 Input Impedance

The whole network impendance including the load R is calculated by

$$Z = \frac{R + j\omega L}{1 + j\omega C(R + j\omega L)} = \frac{R + j\omega L}{1 - \omega^2 CL + j\omega RC} \tag{2.1}$$

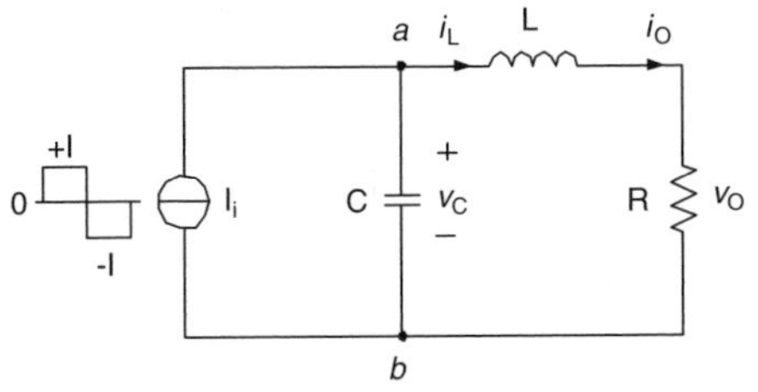

FIGURE 2.9
The equivalent circuit diagram of the two-element RPC.

The natural resonant radian frequency is

$$\omega_0 = \frac{1}{\sqrt{CL}} \tag{2.2}$$

Using the relevant frequency β

$$\beta = \frac{\omega}{\omega_0} = \frac{\omega}{\sqrt{CL}} \tag{2.3}$$

and the quality factor Q

$$Q = \frac{\omega_0 L}{R} = \frac{1}{\omega_0 RC} \tag{2.4}$$

we rewrite the input impendance

$$\frac{Z}{R} = \frac{1 + j\beta Q}{1 - \beta^2 + j\beta / Q} = \left|\frac{Z}{R}\right| \angle\phi \tag{2.5}$$

where

$$\left|\frac{Z}{R}\right| = \sqrt{\frac{1 + (\beta Q)^2}{(1 - \beta^2)^2 + (\beta / Q)^2}} \qquad \phi = \tan^{-1}\beta Q - \tan^{-1}\frac{\beta / Q}{1 - \beta^2}$$

2.3.2 Current Transfer Gain

The current transfer gain is calculated by

$$G(\omega) = \frac{i_O}{i_{in}} = \frac{1}{1 + j\omega C(R + j\omega L)} = \frac{1}{1 - \omega^2 CL + j\omega RC} \tag{2.6}$$

Defining an auxiliary parameter B

$$B(\omega) = 1 + j\omega C(R + j\omega L) \tag{2.7}$$

Hence,

$$G(\omega) = \frac{1}{B(\omega)} = |G| \angle\theta \tag{2.8}$$

Using the relevant frequency and quality factor,

$$G(\beta) = \frac{i_O}{i_{in}} = \frac{1}{1-\beta^2 + j\dfrac{\beta}{Q}} \tag{2.9}$$

$$B(\beta) = 1-\beta^2 + j\frac{\beta}{Q} \tag{2.10}$$

and

$$|G| = \sqrt{\frac{1}{(1-\beta^2)^2 + (\beta/Q)^2}} \qquad \theta = -\tan^{-1}\frac{\beta/Q}{1-\beta^2}$$

2.3.3 Operation Analysis

Based on the equivalent circuit in Figure 2.9, the state equation is established as:

$$\begin{pmatrix} \dot{i}_L \\ \dot{v}_C \end{pmatrix} = \begin{pmatrix} -\dfrac{R}{L} & \dfrac{1}{L} \\ -\dfrac{1}{C} & 0 \end{pmatrix} \begin{pmatrix} i_L \\ v_C \end{pmatrix} + \begin{pmatrix} 0 \\ \dfrac{I}{C} \end{pmatrix} \tag{2.11}$$

By Laplace transform, the state equation in time domain could be transferred into the *s*-domain, given by:

$$\begin{cases} sLI_L(s) - LI_L(0) = V_C(s) - I_L(s)R \\ sCV_C(s) - CV_C(0) = I/s - I_L(s) \end{cases} \tag{2.12}$$

yielding:

$$I_L(s) = \frac{sLCI_L(0) + I/s + CV_C(0)}{s^2LC + sRC + 1} = \frac{s[I_L(0) - I] + [V_C(0) - IR]/L}{(s + \dfrac{R}{2L})^2 + \dfrac{1}{LC} - \dfrac{R^2}{4L^2}} + \frac{I}{s} \tag{2.13}$$

The inductor current in the time-domain is then derived by taking the inverse Laplace transform, giving:

$$i_L(t) = a_1 e^{-\alpha t}\cos\lambda t + \frac{b_1 - a_1\alpha}{\lambda} e^{-\alpha t}\sin\lambda t + I \tag{2.14}$$

where

$$a_1 = I_L(0) - I \qquad b_1 = [V_C(0) - IR]/L$$

α is the damping ratio, $\alpha = R/2L$. λ is the resonant radian frequency, $\lambda = \sqrt{(1/LC) - (R^2/4L^2)}$.

Similarly, the resonant capacitor voltage in the s-domain is attained as:

$$\begin{aligned} V_C(s) &= \frac{sLCV_C(0) + \frac{IR}{s} + [L(I_L(0) - I) + RCV_C(0)]}{s^2LC + sRC + 1} \\ &= \frac{sLC[V_C(0) - IR] + \{L[I - I_L(0)] + RCV_C(0) - IR^2C\}}{s^2LC + sRC + 1} + \frac{IR}{s} \end{aligned} \tag{2.15}$$

The corresponding expression in the time-domain is written by:

$$v_C(t) = a_2 e^{-\alpha t}\cos\lambda t + \frac{b_2 - a_2\alpha}{\lambda} e^{-\alpha t}\sin\lambda t + IR \tag{2.16}$$

where

$$a_2 = V_C(0) - IR \qquad b_2 = \{L[I - I_L(0)] + RCV_C(0) - IR^2C\}/LC$$

To make the analytic Equation (2.14) and Equation (2.16) available, the initial conditions must be known at first. Here the periodic nature is applied under steady state operation. Namely, during one switching cycle, the resonant voltage and current at the initial instant should be the same absolute values with negative sign as those at the half cycle, that is

$$i_L(0) = I_L(0) = -i_L(\frac{T_s}{2}) \qquad v_C(0) = V_C(0) = -v_C(\frac{T_s}{2}) \tag{2.17}$$

Substituting Equation (2.17) into Equation (2.14) and Equation (2.16), respectively, yields:

$$\begin{cases} -I_L(0) = a_1 e^{-\frac{\alpha T_s}{2}}\cos(\frac{\lambda T_s}{2}) + \frac{b_1 - a_1\alpha}{\lambda} e^{-\frac{\alpha T_s}{2}}\sin(\frac{\lambda T_s}{2}) + I \\ -V_C(0) = a_2 e^{-\frac{\alpha T_s}{2}}\cos(\frac{\lambda T_s}{2}) + \frac{b_2 - a_2\alpha}{\lambda} e^{-\frac{\alpha T_s}{2}}\sin(\frac{\lambda T_s}{2}) + IR \end{cases} \tag{2.18}$$

Rearranging,

$$\begin{cases} m_{11}V_C(0) + m_{12}I_L(0) = n_1 \\ m_{21}V_C(0) + m_{22}I_L(0) = n_2 \end{cases} \tag{2.19}$$

Where

$$m_{11} = \frac{1}{L}e^{-\frac{\alpha T_s}{2}}\sin(\frac{\lambda T_s}{2})$$

$$m_{12} = e^{-\frac{\alpha T_s}{2}}[\lambda\cos(\frac{\lambda T_s}{2}) - \alpha\sin(\frac{\lambda T_s}{2})] + \lambda$$

$$n_1 = Ie^{-\frac{\alpha T_s}{2}}[\lambda\cos(\frac{\lambda T_s}{2}) + \alpha\sin(\frac{\lambda T_s}{2})] - \lambda I$$

$$m_{21} = e^{-\frac{\alpha T_s}{2}}[\lambda\cos(\frac{\lambda T_s}{2}) + \alpha\sin(\frac{\lambda T_s}{2})] + \lambda$$

$$m_{22} = -\frac{1}{C}e^{-\frac{\alpha T_s}{2}}\sin(\frac{\lambda T_s}{2})$$

$$n_2 = Ie^{-\frac{\alpha T_s}{2}}[R\lambda\cos(\frac{\lambda T_s}{2}) - \frac{1-\alpha RC}{C}\sin(\frac{\lambda T_s}{2})] - \lambda IR$$

Since all the parameters shown in the coefficients of Equation (2.19) are constant for a given circuit, the initial values of the resonant voltage and current can be calculated. Thus, the complete expressions of the resonant voltage and current are obtained by substituting $V_C(0)$ and $I_L(0)$ into Equation (2.14) and Equation (2.16).

Figure 2.10 shows the general waveforms of resonant voltage $v_C(t)$ and current $i_L(t)$. As seen, both of them are oscillatory and track-following. Actually, further investigation of Equation (2.14) states that inductor current can be rewritten as:

$$i_L(t) = a_1e^{-\alpha t}\cos\lambda t + \frac{b_1 - a_1\alpha}{\lambda}e^{-\alpha t}\sin\lambda t + I = Ae^{-\alpha t}\sin(\lambda t + \varphi) + I \tag{2.20}$$

where

$$A = \sqrt{a_1^2 + (\frac{b_1 - a_1\alpha}{\lambda})^2} \qquad \varphi = \tan^{-1}(\frac{a_1\lambda}{b_1 - a_1\alpha})$$

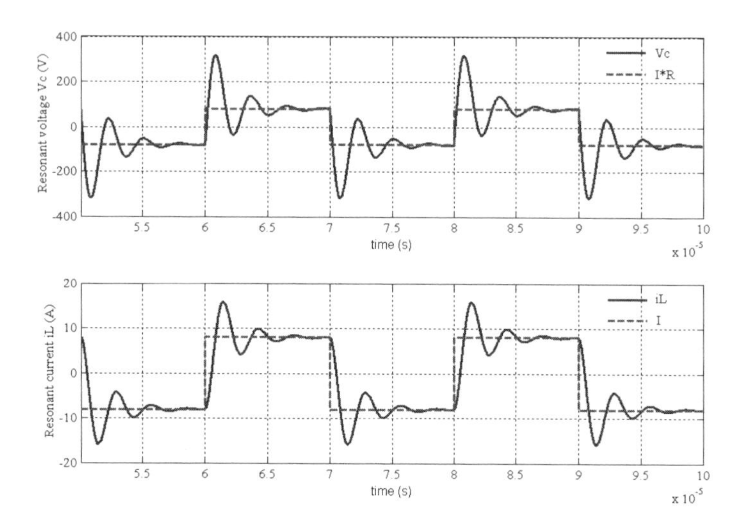

FIGURE 2.10
General waveforms of the resonant voltage v_C and current i_L.

It is apparent that the inductor current is composed of two different components. One is the oscillatory component, consisting of the sinusoidal function. The other is the compulsory component, given as the input current I. The oscillatory component is time-attenuation. With the time increasing, it will be attenuated to zero and the inductor current is then convergent to the compulsory component. The attenuating rate is determined by the damping ratio α. Larger ratios cause faster attenuation. Similar conclusions can also be made about the capacitor voltage $v_C(t)$.

When the value of damping factor $e^{-\alpha T_s/2}$ is approaching 1, the resonant voltage and current will contain an undamped oscillatory component. Hence, the coefficients in Equation (2.19) can be simplified accordingly, giving:

$$m_{11}^{'} = \frac{1}{L}\sin(\frac{\lambda T_s}{2})$$

$$m_{12}^{'} = \lambda\cos(\frac{\lambda T_s}{2}) - \alpha\sin(\frac{\lambda T_s}{2}) + \lambda$$

$$n_1^{'} = \lambda[\cos(\frac{\lambda T_s}{2}) - 1]I + \alpha\sin(\frac{\lambda T_s}{2})I$$

$$m_{21}^{'} = \lambda\cos(\frac{\lambda T_s}{2}) + \alpha\sin(\frac{\lambda T_s}{2}) + \lambda$$

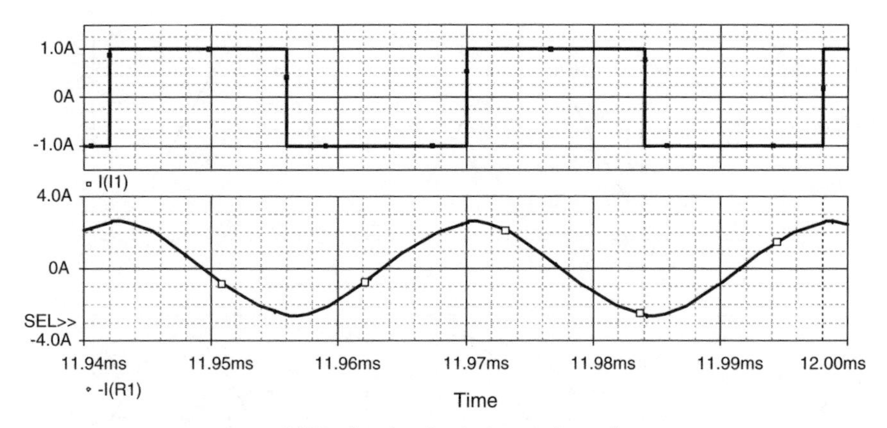

(a) The input and output current waveforms

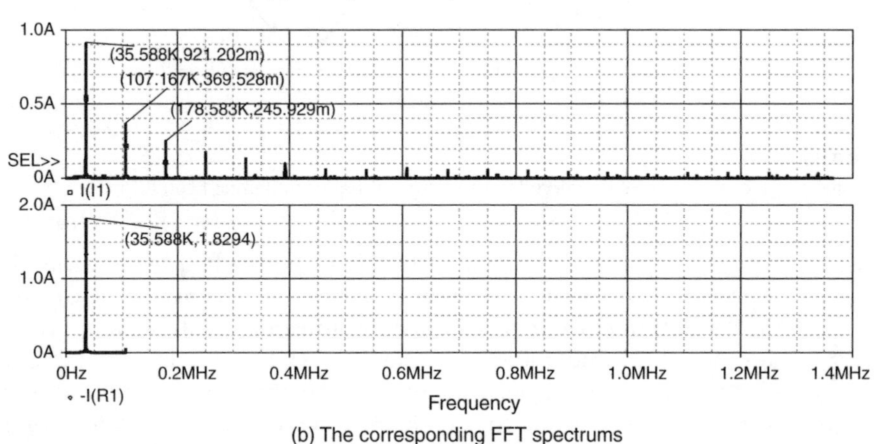

(b) The corresponding FFT spectrums

FIGURE 2.11
The simulation waveforms of the input and output currents.

$$m_{22}^{'} = -\frac{1}{C}\sin(\frac{\lambda T_s}{2})$$

$$n_2^{'} = I[R\lambda\cos(\frac{\lambda T_s}{2}) - \frac{1-\alpha RC}{C}\sin(\frac{\lambda T_s}{2})] - \lambda IR$$

2.3.4 Simulation Results

For the purpose of verifying the mathematical derivations, a prototype bipolar current source resonant inverter is proposed for simulation and experiments. The simulation is carried out by Pspice and the results are presented in Figure 2.11a, where the upper channel is the input square-wave current I_i

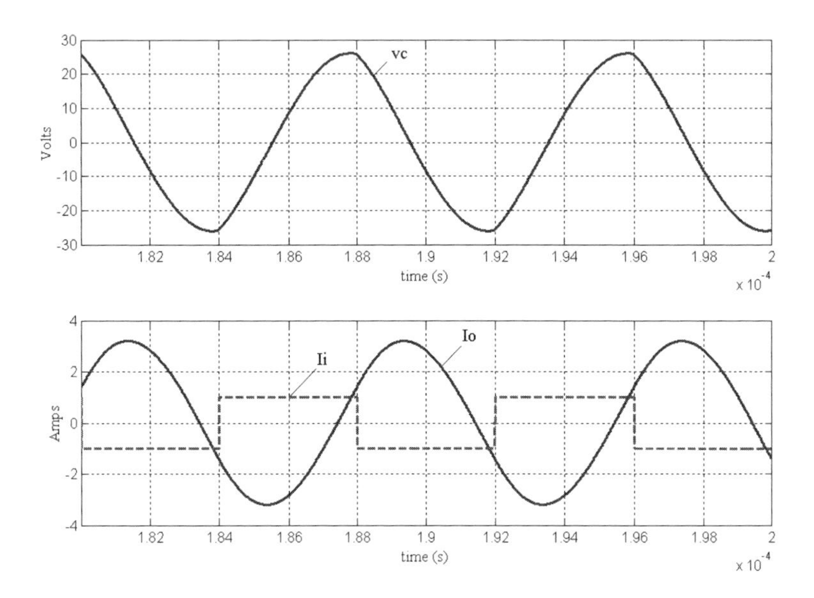

FIGURE 2.12
Tested waveforms of the resonant voltage and current under undamped condition.

and the lower channel is the output current i_o, respectively. The corresponding FFT spectrums are shown in Figure 2.11b. From the results, it can be found that the output current is very sleekly sinusoidal with the amplitude larger than the input fundamental current. The THD value is only 1.2%. It is obvious that the resonant network has the capability of attenuating the higher order harmonics in the input square-wave current and transferring their energy into the output current. Due to the advantages of high current transfer gain and negligible output harmonics, this inverter could be widely used in many high-frequency huge-current applications.

2.3.5 Experimental Results

The corresponding resonant waveforms under this condition are shown in Figure 2.12, where both the capacitor voltage and the inductor current, i.e., output current are undamped oscillatory.

Bibliography

Ang, S.S., *Power Switching Converters*, New York: Marcel Dekker, 1995.

Batarseh, I., Resonant converter topologies with three and four energy storage elements, *IEEE Transactions on Power Electronics*, 9, 64, 1994.

Belaguli, V. and Bhat, A.K. S., Series-parallel resonant converter operating in discontinuous current mode-analysis, design, simulation, and experimental results, *IEEE Transactions on Circuits and System*, 47, 433, 2000.

Bhat, A.K.S., Analysis and design of a series-parallel resonant converter with capacitive output filter, *IEEE Transactions on Industry Applications*, 27, 523, 1991.

Bhat, A.K.S. and Dewan, S.B., Analysis and design of a high frequency resonant converter using LCC type Commutation, in *Proceedings of IEEE Industry Applications Society Annual Meeting*, 1986, p. 657.

Cosby, Jr., M.C. and Nelms, R.M., A resonant inverter for electronic ballast applications, *IEEE Transactions on Industry Electronics*, 41, 418, 1994.

Johnson, S.D. and Erickson, R.W., Steady-state analysis and design of the parallel resonant converter, *IEEE Transactions on Power Electronics*, 3, 93, 1988.

Jones, C.B. and Vergez, J.P., Application of PWM techniques to realize a 2 MHz offline switching regulator, with hybrid implementations, in *Proceedings of IEEE APEC'87*, 1986, p. 221.

Kang, Y.G. and Upadhyay, A.K., Analysis and design of a half-bridge parallel resonant converter operating above resonance, *IEEE Transactions on Industry Applications*, 27, 386, 1991.

King, R.J. and Stuart, T.A., Modeling the full-bridge series-resonant power converter, *IEEE Transaction on Aerospace and Electronic System*, 18, 449, 1982.

Kisch, J.J. and Perusse, E.T., Megahertz power converters for specific power systems, in *Proceedings of IEEE APEC'87*, 1987, p. 115.

Kislovski, A.S., Redl, R., and Sokal, N.O., *Dynamic Analysis of Switching-Mode DC/DC Converters*, New York, Van Nostrand Reinhold, 1991.

Mitchell, D.M., *DC-DC Switching Regulator Analysis*, New York, McGraw-Hill, 1988.

Nathan, B.S. and Ramanarayanan, V., Analysis, simulation and design of series resonant converter for high voltage applications, in *Proceedings of IEEE International Conference on Industrial Technology'00*, 2000, p. 688.

Severns, R.P., Topologies for three-element resonant converters, *IEEE Transactions on Power Electronics*, 7, 89, 1992.

Steigerwald, R.L., A comparison of half-bridge resonant converter topologies, *IEEE Transactions on Power Electronics*, 3, 174, 1988.

Tanaka, J., Yuzurihara I., and Watanabe, T., Analysis of a full-bridge parallel resonant converter, in *Proceedings of 13th International Telecommunications Energy Conference*, 1991, p. 302.

Witulski, A. and Erickson, R.W., Steady-state analysis of the series resonant converter, *IEEE Transactions on Aerospace Electronics*, 21, 791, 1985.

3

Π-CLL Current Source Resonant Inverter

3.1 Introduction

This chapter introduces a three-element current source resonant inverter (CSRI): Π-CLL CSRI consists of three energy-storage elements CLL, the Π-CLL. A bipolar current source is employed in this circuit as described in the previous chapter and shown in Figure 2.7. Since there is no transformer and the circuit is not working in push-pull operation, the control circuit is very simple and power losses are low. This Π-CLL CSRI is shown in Figure 3.1. The energy source is a DC voltage V_{in}, which is chopped by two mains switches S_1 and S_2. The three energy-storing elements are C, L_1 and L_2. Two inductors can be different, i.e., $L_2 = p\, L_1$ with p as a random value. The load can be a coil, transformer or HF annealing equipment. A resistance R_{eq} is assumed. Its equivalent circuit diagram is shown in Figure 3.2.

3.1.1 Pump Circuits

In this application two boost pumps are used working at the conduction duty $k = 0.5$. Each pump consists of the common DC voltage source V_{in}, a switch S, and a large inductor L. The pump-out energy is usually measured by output current injection.

3.1.2 Current Source

An ideal current source has infinite impedance and constant output current. If the inductance of the pump circuits is large enough, the current flowing through it may keep a nearly constant value. The internal equivalent impedance of the pump is very high. The ripple of the inductor current depends on the input voltage, inductor's inductance, switching frequency, and conduction duty.

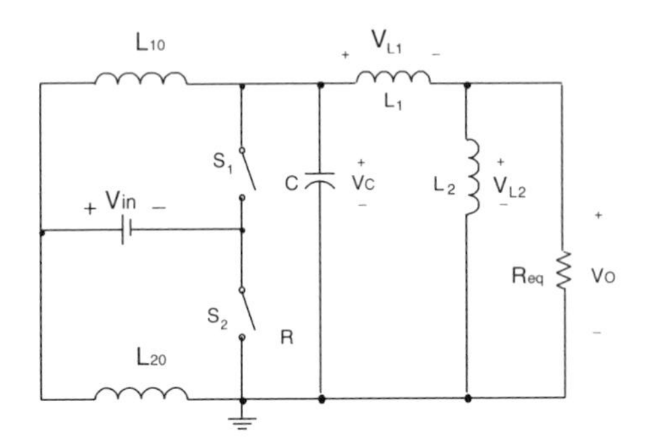

FIGURE 3.1
The Π-CLL current source resonant inverter.

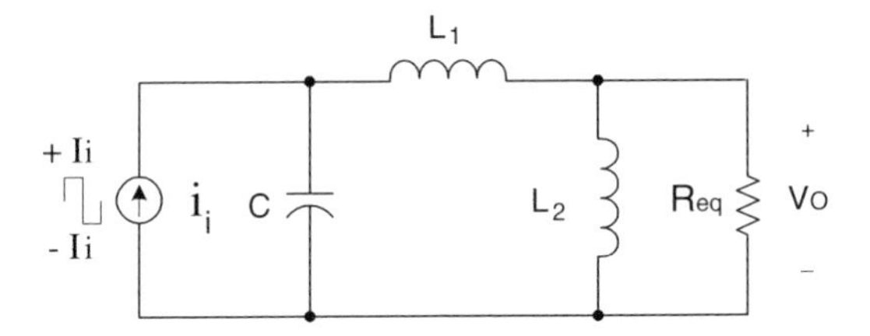

FIGURE 3.2
The equivalent circuit.

3.1.3 Resonant Circuit

Like other resonant inverters, this Π-CLL CSRI consists of a resonant circuit sandwiched between the current source (input switching circuit) and the output load. The resonant circuit can be considered as a Π-CLL low-pass filter. By network theory, these three components construct a circuit that is no longer a real low-pass filter. Therefore, the Π-CLL circuit has two peak resonant points. It gives more convenience to designers to match in the industrial applications.

3.1.4 Load

Load is usually resistive load or inductive plus resistive load. To simplify the problem we use R_{eq} to represent either an actual or an equivalent resistance, to consume the output power. This assumption is reasonable, because the load inductance can be considered the parallel part to the inductor L_2.

3.1.5 Summary

The switching circuit consists of two pumps (two boost pumps employed): the R_{eq} is as mentioned, either an actual or an equivalent AC load resistance.

The source current i_i is a constant current yielded by input voltage V_i via large inductors L_{10} and L_{20}. To operate this circuit, S_1 is turned on and off in 180° (S_2 is idle) at the frequency $\omega = 2\pi f$. After the switching circuit, input current can be considered a bipolar square-wave current alternating in value between $+I_i$ and $-I_i$ that is then input to the resonant circuit section. The equivalent circuit is shown in Figure 3.2.

3.2 Mathematic Analysis

In order to concentrate the function analysis of this Π-CLL, we assume:

1. The inverter's source is a constant current source determined by the pump circuits.
2. Two MOSFETs in the switching circuit are turned-on and turned-off 180° out of phase with each other at same switching frequency and with a duty cycle of 50%.
3. Two switches are ideal components without on-resistance, and negligible parasitic capacitance and zero switching time.
4. Two diodes are components having a zero forward voltage drop and forward resistance.
5. Four energy-storage elements are passive, linear, time-invariant, and do not have parasitic reactive components.

Using these assumptions, the following analyses are based on using Figure 3.2.

3.2.1 Input Impedance

This CSRI is a third-order system. The mathematic analysis of operation and stability is more complex than three energy-storage element current source resonant inverters. The input impedance is given by

$$Z(\omega) = \frac{-\omega^2 L_1 L_2 + j\omega R_{eq}(L_1 + L_2)}{R_{eq}[1 - \omega^2 (L_1 + L_2)C] + j\omega L_2 (1 - \omega^2 L_1 C)} \tag{3.1}$$

The corresponding phase angle is

$$\phi(\omega) = \tan^{-1}\{\frac{\omega R_{eq}(L_1 + L_2)}{-\omega^2 L_1 L_2}\} - \tan^{-1}\{\frac{\omega L_2 (1 - \omega^2 L_1 C)}{R_{eq}[1 - \omega^2 (L_1 + L_2)C]}\} \tag{3.2}$$

Define

$$B(\omega) = R_{eq}[1-\omega^2(L_1+L_2)C] + j\omega L_2(1-\omega^2 L_1 C) \quad (3.3)$$

So that

$$Z(\omega) = \frac{-\omega^2 L_1 L_2 + j\omega R_{eq}(L_1+L_2)}{B(\omega)} \quad (3.4)$$

3.2.2 Components' Voltages and Currents

This Π-CLL CSRI has three resonant components C, L_1, and L_2, and the output equivalent resistance R_{eq}. In order to compare with the parameters easily, all components' voltages and currents are responded to the input fundamental current I_i. All transfer functions are in the frequency domain (or the ω-domain).

Voltage and current on capacitor C:

$$\frac{V_C(\omega)}{I_{in}(\omega)} = Z(\omega) = \frac{-\omega^2 L_1 L_2 + j\omega R_{eq}(L_1+L_2)}{B(\omega)} \quad (3.5)$$

$$\frac{I_C(\omega)}{I_i(\omega)} = \frac{-\omega^2 R_{eq} C(L_1+L_2) - j\omega^3 C L_1 L_2}{B(\omega)} \quad (3.6)$$

Voltage and current on inductor L_1:

$$\frac{V_{L1}(\omega)}{I_{in}(\omega)} = \frac{-\omega^2 L_1 L_2 + j\omega R_{eq} L_1}{B(\omega)} \quad (3.7)$$

$$\frac{I_{L1}(\omega)}{I_{in}(\omega)} = \frac{R_{eq} + j\omega L_2}{B(\omega)} \quad (3.8)$$

Voltage and current on inductor L_2:

$$\frac{V_{L2}(\omega)}{I_{in}(\omega)} = \frac{j\omega R_{eq} L_2}{B(\omega)} \quad (3.9)$$

$$\frac{I_{L2}(\omega)}{I_{in}(\omega)} = \frac{R_{eq}}{B(\omega)} \quad (3.10)$$

The output voltage and current on the resistor R_{eq}:

$$\frac{V_O(\omega)}{I_{in}(\omega)} = \frac{V_{L2}(\omega)}{I_{in}(\omega)} = \frac{j\omega R_{eq} L_2}{B(\omega)} \tag{3.11}$$

The current transfer gain is given by

$$\begin{aligned} g(\omega) &= \frac{I_O(\omega)}{I_{in}(\omega)} = \frac{j\omega L_2}{B(\omega)} = \frac{j\omega L_2}{R_{eq}[1-\omega^2(L_1+L_2)C] + j\omega L_2(1-\omega^2 L_1 C)} \\ &= \frac{1}{(1-\omega^2 L_1 C) - j\dfrac{R_{eq}[1-\omega^2(L_1+L_2)C]}{\omega L_2}} = |g|\angle\theta \end{aligned} \tag{3.12}$$

Thus,

$$|g(\omega)| = \frac{1}{\sqrt{(1-\omega^2 L_1 C)^2 + \dfrac{R_{eq}{}^2[1-\omega^2(L_1+L_2)C]^2}{(\omega L_2)^2}}} \tag{3.13}$$

$$\theta(\omega) = \tan^{-1}\frac{R_{eq}[1-\omega^2(L_1+L_2)C]}{\omega L_2(1-\omega^2 L_1 C)} \tag{3.14}$$

3.2.3 Simplified Impedance and Current Gain

Usually, we are interested in the input impedance and output current gain rather than all transfer functions listed in previous section. To simplify the operation, we can select:

$$\omega_0 = \frac{1}{\sqrt{L_1 C}} \qquad \beta = \frac{\omega}{\omega_0} \qquad Q = \frac{\omega_0 L}{R_{eq}} = \frac{1}{\omega_0 C R_{eq}} \qquad \text{and} \qquad p = \frac{L_2}{L_1}$$

Hence

$$\begin{aligned} Z(\omega) &= \frac{-\beta^2\omega_0{}^2 L_1 L_2 + j\beta\omega_0 R_{eq}(L_1+L_2)}{R_{eq}[1-\beta^2\omega_0{}^2(L_1+L_2)C] + j\beta\omega_0 L_2(1-\beta^2\omega_0{}^2 L_1 C)} \\ &= \frac{-\beta^2\omega_0{}^2 p L_1{}^2 + j\beta\omega_0 R_{eq} L_1(1+p)}{R_{eq}[1-\beta^2\omega_0{}^2 L_1(1+p)C] + j\beta\omega_0 p L_1(1-\beta^2\omega_0{}^2 L_1 C)} \\ &= \frac{-p(\beta Q R_{eq})^2 + j(1+p)\beta Q R_{eq}{}^2}{R_{eq}\{[1-(1+p)\beta^2] + jp\beta Q(1-\beta^2)\}} = |Z|\angle\phi \end{aligned} \tag{3.15}$$

Then obtain,

$$B(\beta) = R_{eq}[1-(1+p)\beta^2 + jp\beta Q(1-\beta^2)] \tag{3.16}$$

and

$$\begin{aligned}\phi(\omega) &= \tan^{-1}\{\frac{\beta\omega_0 R_{eq}L_1(1+p)}{-p\beta^2\omega_0{}^2L_1{}^2}\} - \tan^{-1}\{\frac{p\beta\omega_0 L_1(1-\beta^2\omega_0{}^2L_1C)}{R_{eq}[1-\beta^2\omega_0{}^2L_1(1+p)C]}\} \\ &= \tan^{-1}\frac{1+p}{-p\beta Q} - \tan^{-1}\frac{p\beta Q(1-\beta^2)}{1-\beta^2(1+p)} \\ &= \pi - \tan^{-1}\frac{1+p}{p\beta Q} - \tan^{-1}\frac{p\beta Q(1-\beta^2)}{1-\beta^2(1+p)}\end{aligned} \tag{3.17}$$

Therefore,

$$Z = \frac{-p\beta^2Q^2 + j\beta Q(1+p)}{1-(1+p)\beta^2 + jp\beta Q(1-\beta^2)}R_{eq} = |Z| < \phi \tag{3.18}$$

where

$$|Z| = \frac{\sqrt{p^2\beta^4Q^4 + \beta^2Q^2(1+p)^2}}{\sqrt{[1-(1+p)\beta^2]^2 + [p\beta Q(1-\beta^2)]^2}}R_{eq}$$

$$\frac{|Z|}{R_{eq}} = \frac{\sqrt{(p\beta^2Q^2)^2 + [\beta Q(1+p)]^2}}{\sqrt{[1-(1+p)\beta^2]^2 + [p\beta Q(1-\beta^2)]^2}} \tag{3.19}$$

and

$$\phi = \pi - \tan^{-1}\frac{1+p}{p\beta Q} - \tan^{-1}\frac{p\beta Q(1-\beta^2)}{1-(1+p)\beta^2}$$

The characteristics of input impedance $|Z|/R_{eq}$ vs. relevant frequency β referring to $p = 0.5$ and various Q, is shown in Figure 3.3 and Table 3.1.

The characteristics of phase angle ϕ vs. relevant frequency β referring to $p = 0.5$ and various Q, is shown in Figure 3.4 and Table 3.2.

The current transfer gain becomes

$$g = \frac{1}{\frac{1-(1+p)\beta^2}{jpQ} + 1 - \beta^2} = |g| < \theta \tag{3.20}$$

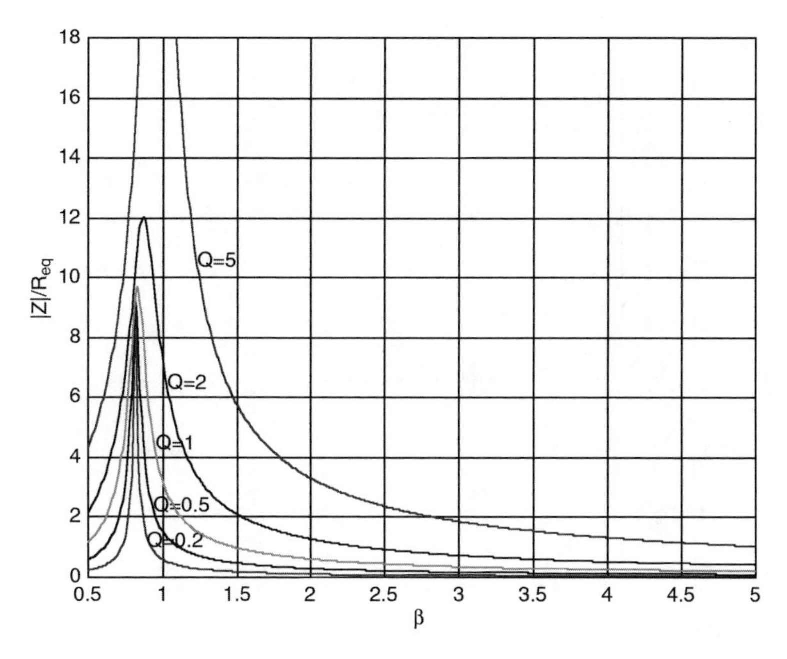

FIGURE 3.3
The curves of $|Z|/R_{eq}$ vs. β referring to Q.

TABLE 3.1
$|Z|/R_{eq}$ vs. β Referring to p = 0.5 and Various Q

	β = 0.5	**0.678**	**0.755**	**0.83**	**1.0**	**1.5**	**2.0**	**3.0**	**5.0**
Q = 0.2	0.2397	0.6513	1.5266	5.9126	0.6013	0.1898	0.1202	0.0721	0.0412
0.5	0.5954	1.5809	3.4353	8.6490	1.5207	0.4790	0.3029	0.1814	0.1033
1	1.1652	2.8924	5.3662	9.6875	3.1623	0.9852	0.6183	0.3673	0.2076
2	2.1693	4.6483	7.1323	10.9302	7.2111	2.1031	1.2804	0.7437	0.4162
5	4.3323	7.9346	11.0396	16.4384	29.1548	5.7645	3.3015	1.8719	1.0415

where

$$|g| = \frac{1}{\sqrt{[\frac{1-(1+p)\beta^2}{pQ}]^2 + (1-\beta^2)^2}} \tag{3.21}$$

and

$$\theta = -\tan^{-1}\frac{1-(1+p)\beta^2}{p(1-\beta^2)Q}$$

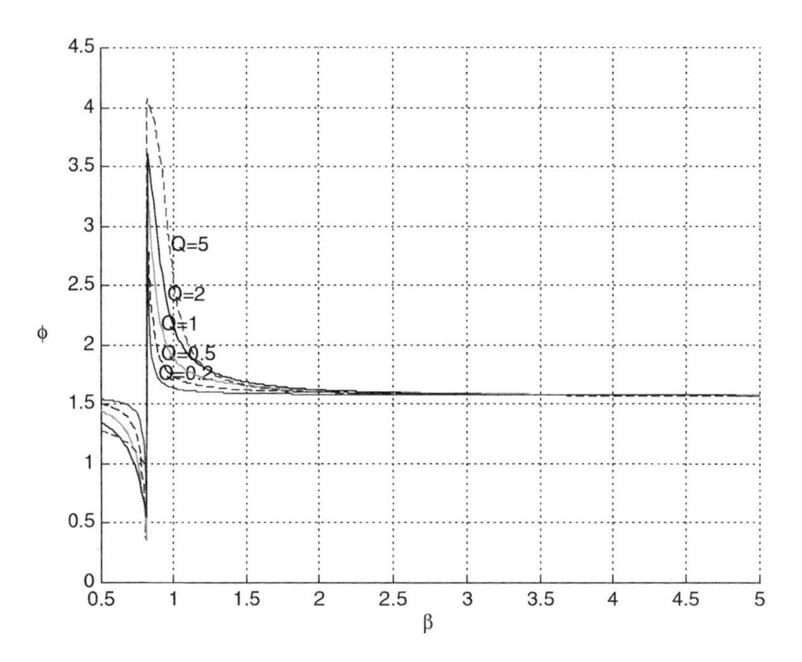

FIGURE 3.4
The curves of ϕ vs. β referring to Q.

TABLE 3.2
ϕ vs. β Referring to p = 0.5 and Various Q

	β = 0.5	0.678	0.755	0.83	1.0	1.5	2.0	3.0	5.0
Q = 0.2	1.5442	1.4985	1.4008	2.2849	1.6374	1.5917	1.5839	1.5785	1.5749
0.5	1.5050	1.3965	1.1856	2.8022	1.7359	1.6209	1.6011	1.5869	1.5776
1	1.4445	1.2601	0.9755	3.1587	1.8925	1.6585	1.6184	1.5912	1.5769
2	1.3521	1.1276	0.8863	3.5185	2.1588	1.6879	1.6220	1.5873	1.5746
5	1.2827	1.1732	1.0760	4.0348	2.6012	1.6593	1.6011	1.5788	1.5724

The characteristics of current transfer gain $|g|$ vs. relevant frequency β referring to p = 0.5, 1 and 2, and various Q, is shown in Figure 3.5 to Figure 3.7 and Table 3.3 to Table 3.5.

The characteristics of the phase angle θ vs. relevant frequency β referring to p = 0.5 and various Q, is shown in Figure 3.8 and Table 3.6.

For various β and Q we have different current transfer gain. For example, when $\beta^2 = 1$ with any p, we have $g = -jQ = Q \angle 90°$. It means that the output current can be larger than the fundamental harmonic of the input current! The larger the value of Q is, the higher the gain g.

Actually, set $\beta^2 = t$, we can find the maximum $|g|$ from

$$\frac{d}{dt}|g| = 0$$

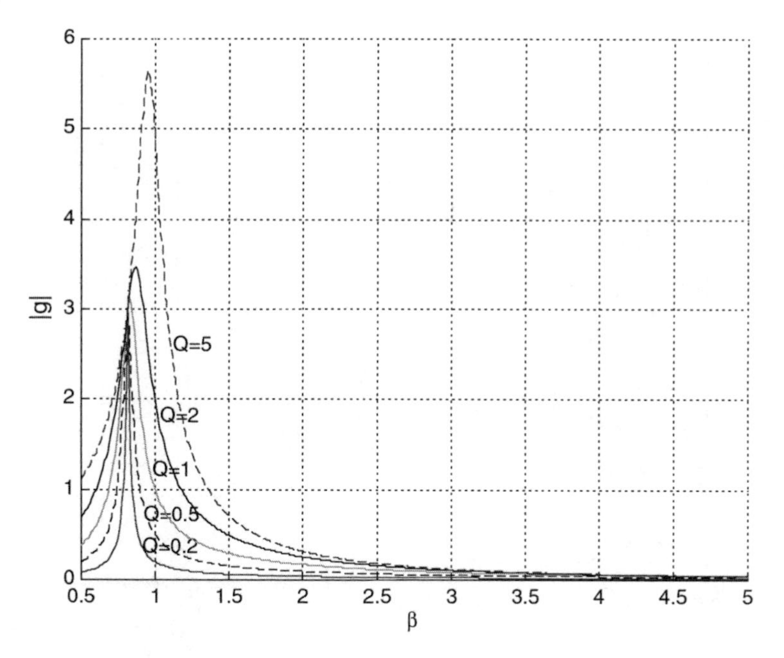

FIGURE 3.5
The curves of $|g|$ vs. β referring to Q, $p = 0.5$.

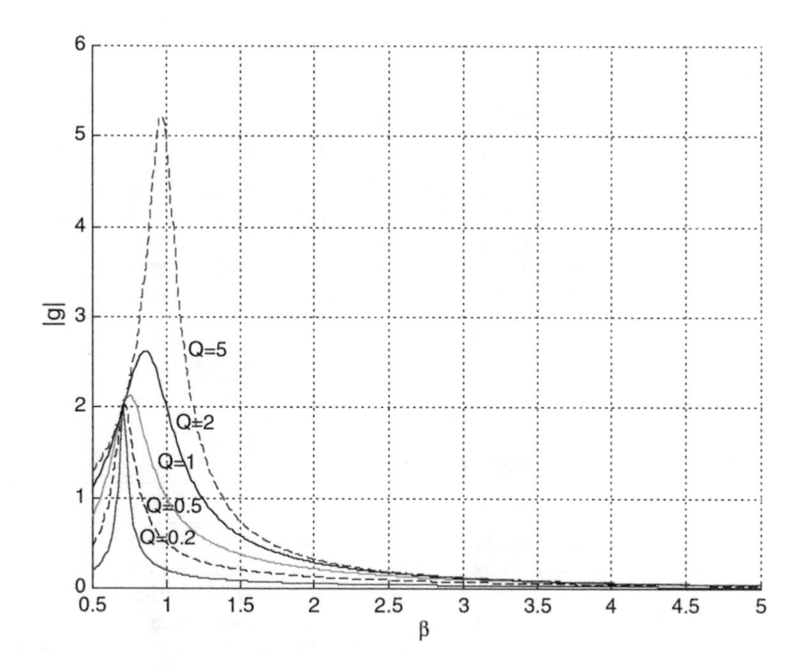

FIGURE 3.6
The curves of $|g|$ vs. β referring to Q, $p = 1$.

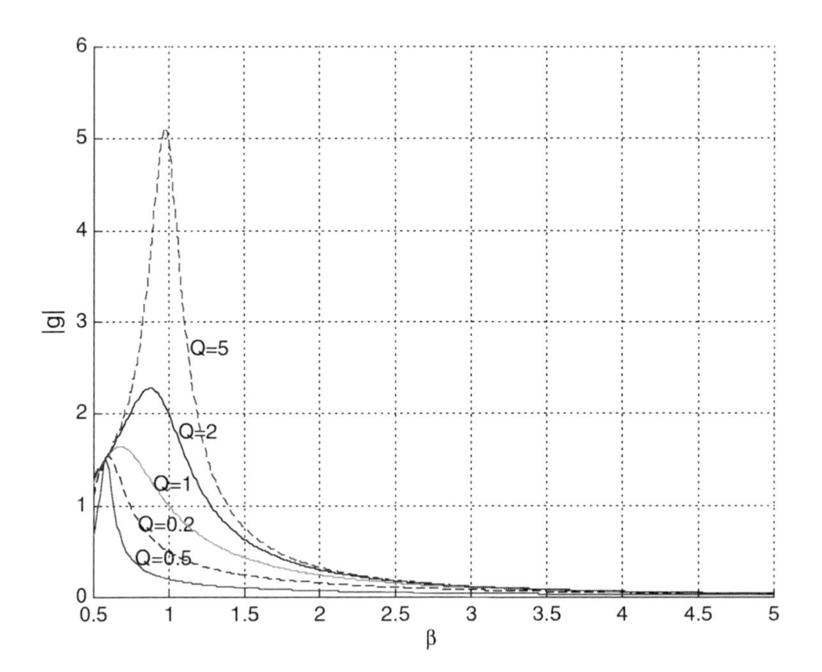

FIGURE 3.7
The curves of $|g|$ vs. β referring to Q, $p = 2$.

TABLE 3.3
$|g|$ vs. β Referring to $p = 0.5$ and Various Q

	$\beta = 0.5$	0.678	0.755	0.83	1.0	1.5	2.0	3.0	5.0
$Q = 0.2$	0.0799	0.2169	0.5082	1.9679	0.2000	0.0630	0.0397	0.0236	0.0130
0.5	0.1978	0.5236	1.1361	2.8558	0.5000	0.1549	0.0958	0.0541	0.0265
1	0.3831	0.9404	1.7346	3.1122	1.0000	0.2937	0.1715	0.0866	0.0356
2	0.6860	1.4119	2.1236	3.1879	2.0000	0.4957	0.2561	0.1109	0.0399
5	1.1094	1.7528	2.2895	3.2101	5.0000	0.7136	0.3162	0.1224	0.0414

TABLE 3.4
$|g|$ vs. β Referring to $p = 1$ and Various Q

	$\beta = 0.5$	0.678	0.755	0.83	1.0	1.5	2.0	3.0	5.0
$Q = 0.2$	0.1978	1.2446	0.9782	0.4353	0.2000	0.0852	0.0563	0.0340	0.0183
0.5	0.4682	1,6939	1.7609	1.0394	0.5000	0.2070	0.1313	0.0721	0.0323
1	0.8000	1.8075	2.1355	1.8138	1.0000	0.3778	0.2169	0.1020	0.0386
2	1.1094	1.8397	2.2734	2.5943	2.0000	0.5848	0.2879	0.1178	0.0408
5	1.2883	1.8490	2.3171	3.0850	5.0000	0.7495	0.3246	0.1238	0.0415

or

TABLE 3.5
$|g|$ vs. β Referring to p = 2 and Various Q

	β = 0.5	0.678	0.755	0.83	1.0	1.5	2.0	3.0	5.0
Q = 0.2	0.6860	0.6673	0.4184	0.3098	0.2000	0.1035	0.0711	0.0433	0.0227
0.5	1.1094	1.2862	0.9670	0,7563	0.5000	0.2480	0.1596	0.0848	0.0355
1	1.2649	1.6438	1.5694	1.4007	1.0000	0.4370	0.2457	0.1099	0.0398
2	1.3152	1.7918	2.0406	2.2360	2.0000	0.6349	0.3030	0.1207	0.0412
5	1.3304	1.8409	2.2720	2.9709	5.0000	0.7648	0.3279	0.1243	0.0416

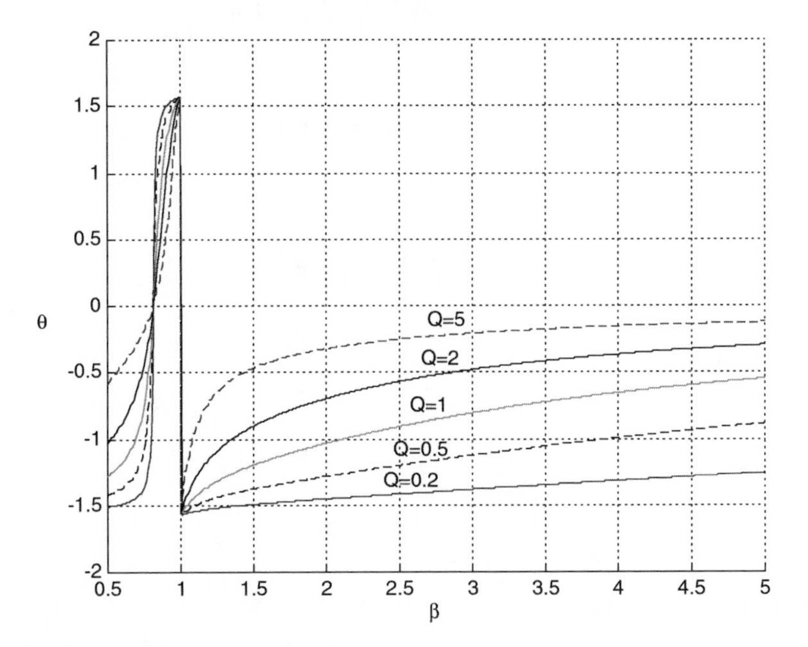

FIGURE 3.8
The curves of θ vs. β referring to Q, p = 0.5.

TABLE 3.6
θ vs. β Referring to p = 0.5 and Various Q

	β = 0.5	0.678	0.755	0.83	1.0	1.5	2.0	3.0	5.0
Q = 0.2	-1.5109	-1.4533	-1.3505	0.9120	1.5708	-1.4920	-1.4514	-1.3811	-1.2532
0.5	-1.4219	-1.2840	-1.0604	0.4769	1.5708	-1.3759	-1.2793	-1.1233	-0.8828
1	-1.2793	-1.0378	-0.7290	0.2528	1.5708	-1.1948	-1.0304	-0.8058	-0.5465
2	-1.0304	-0.7030	-0.4200	0.1284	1.5708	-0.9025	-0.6947	-0.4802	-0.2953
5	-0.5880	-0.3268	-0.1768	0.0516	1.5708	-0.4690	-0.3218	-0.2054	-0.1211

$$\frac{1-(i+p)t}{pQ}(1+p)+(1-t)=0 \tag{3.22}$$

We obtain

$$t=\frac{1+p+pQ}{1+2p+p^2+pQ}=\frac{1}{1+\dfrac{p+p^2}{1+p+pQ}}$$

If taking $Q = 1$, $t = \beta^2 = 0.6$, or $\beta = 0.7746$, $|g| = 2.236$.

If this inverter is working at the conditions: $\beta = 1$ and $Q >> 1$,

$$Z=\frac{jQ}{-1+j\dfrac{1}{Q}}R_{eq} \doteq -jQR_{eq} \tag{3.23}$$

$$\phi=\frac{\pi}{2}-\tan^{-1}\frac{1}{-Q} \doteq \frac{\pi}{2}-\pi=-\frac{\pi}{2}$$

correspondingly

$$g=\frac{1}{-jp/pQ} \doteq jQ \tag{3.24}$$

$$\theta=-\pi/2$$

The Π-CLL circuit is not only the resonant circuit, but the band-pass filter as well. All higher order harmonic components in the input current are effectively filtered by the Π-CLL circuit. The output current is nearly a pure sinusoidal waveform with the fundamental frequency $\omega = 2\pi f$.

3.2.4 Power Transfer Efficiency

The power transfer efficiency is a very important parameter and is calculated here. From Figure 3.5 we know the input current is $i_i(\omega t)$, i.e.,

$$i_i(\omega t)=\begin{cases}+I_i & 2n\pi \le \omega t \le (2n+1)\pi \\ -I_i & (2n+1)\pi \le \omega t \le 2(n+1)\pi\end{cases} \quad \text{with n = 0, 1, 2, 3, ... } \infty \tag{3.25}$$

where $I_i = V_1/Z + j\omega L_{10}$ that varies with different frequency.

This is a square waveform pulse-train. Using fast fourier transform (FFT), we have the spectrum form as

$$i_i(\omega t) = \frac{2I_i}{\pi} \sum_{n=1}^{\infty} \frac{\sin(2n+1)\omega t}{2n+1} \tag{3.26}$$

The fundamental frequency component is

$$i_{fund}(\omega t) = \frac{2I_i}{\pi} \sin \omega t \tag{3.27}$$

Output current is

$$i_2(\omega t) = g\frac{2I_i}{\pi} \sin \omega t \tag{3.28}$$

Because of the assumptions no power losses were considered. The power transfer efficiency from DC source to AC output is calculated in following equations. The total input power is

$$P_{in} = I_i^2 * real(Z) \tag{3.29}$$

The AC output power is gathered in the fundamental component that is

$$P_{fund} = (|g|\frac{2I_i}{\pi\sqrt{2}})^2 R_{eq} \tag{3.30}$$

The power transfer efficiency is

$$\eta = \frac{P_{fund}}{P_{in}} = \frac{\frac{2(I_i|g|)^2}{\pi^2} R_{eq}}{I_i^2 * real(Z)} = \frac{2|g|^2}{\pi^2}\cos\phi \tag{3.31}$$

3.3 Simulation Results

To verify the design and calculation results, PSpice simulation package was applied for these circuits. Choosing $V_I = 30$ V, all pump inductors $L_i = 10$mH, the resonant capacitor $C = 0.2$ μF, and inductors $L_1 = L_2 = 70$ μH, load $R = 10$ Ω, k = 0.5 and $f = 35$ kHz. The input and output current waveforms are

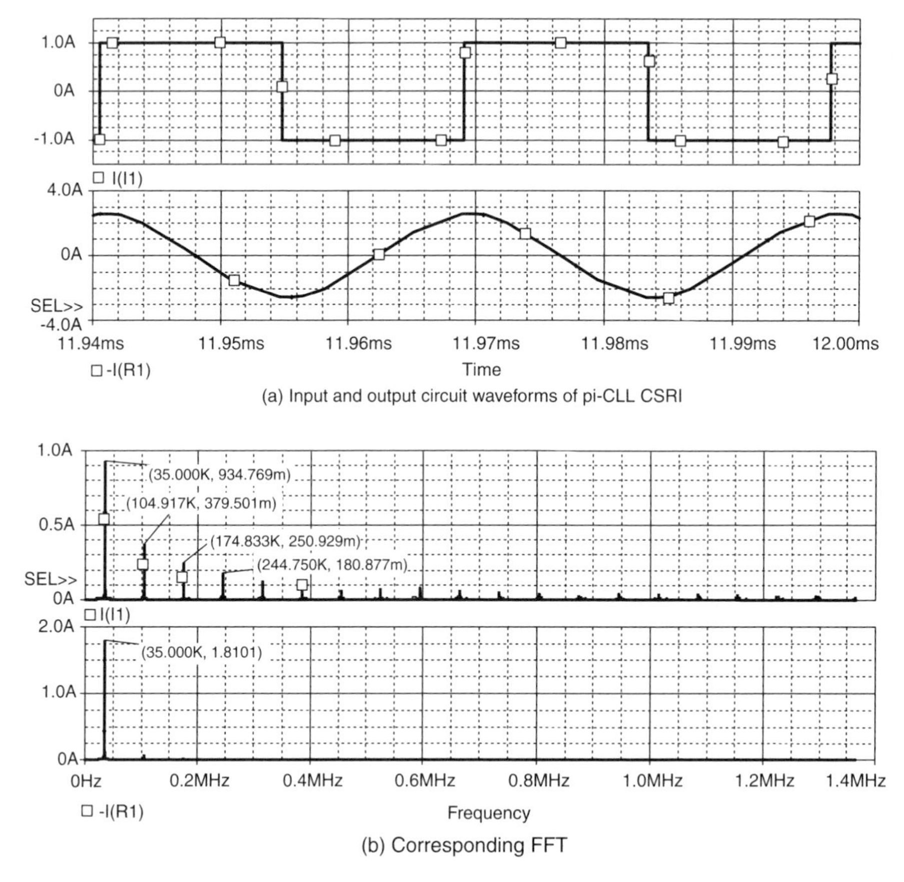

FIGURE 3.9
Input and output current waveforms of Π-CLL circuit at f = 35KHz.

shown in Figure 3.9a. Their corresponding FFT spectrums are shown in Figure 3.9b. It is obviously illustrated that the output waveform is nearly a sinusoidal function, and its corresponding THD is nearly unity.

3.4 Discussion

3.4.1 Function of the Π-CLL Circuit

As a Π-CLC filter, it is a typical low-pass filter. All harmonics with frequency $\omega > \omega_0$ will be blocked. Π-CLL filter circuit has thoroughly different characteristics from that of low-pass filters. It allows the signal with higher frequency $\omega > \omega_0$ (it means $\beta > 1$ in Section 3.2) passing it and enlarging the energy.

3.4.2 Applying Frequency to this Π-CLL CSRI.

From our analysis and verifications we found the fact that the effective applying frequency to this Π-CLL current source resonant inverter is (0.8 to 2.0) f_0. Outside this region both current transfer gain and efficiency are falling fast.

3.4.3 Explanation of $g > 1$

We recognized the fact that current transfer gain is greater than unity from mathematical analysis, and simulation and experimental results. The reason to enlarge the fundamental current is that the resonant circuit transfers the energy of other higher order harmonics to the fundamental component. Therefore the gain of the fundamental current can be greater than unity.

3.4.4 DC Current Component Remaining

Since the Π-CLL resonant circuit could not block the DC component, the output current still remains the same DC current component. This is not useful for most ordinary inverter applications.

3.4.5 Efficiency

From mathematical calculation and analysis, and simulation and experimental results, we can obtain very high efficiency. Its maximum value can be nearly unity! It means this Π-CLL resonant circuit can transfer the energy from not only higher order harmonics, but also a DC component into the fundamental component.

Bibliography

Batarseh, I., Resonant converter topologies with three and four energy storage elements, *IEEE Transactions on PE*, 9, 64, 1994.

Batarseh, I., Liu, R., and Lee, C.Q., State-plan analysis and design of parallel resonant converter with LCC-type commutation, in *Proceedings of IEEE-SICE'88*, Tokyo, 1988, p. 831.

Chen, J. and Bonert, R., Load independent AC/DC power supply for higher frequencies with sinewave output, *IEEE Transactions on IA*, 19, 223, 1983.

Kazimierczuk, M.K. and Cravens, R.C., Current-source parallel-resonant DC/AC inverter with transformer, *IEEE Transactions on PE*, 11, 275, 1996.

Kazimierczuk, M.K. and Czarkowski, D., *Resonant Power Converters*, John Wiley, New York, 1995.

Liu, R., Batarseh, I., and Lee, C.Q., Comparison of performance characteristics between LLC-type and conversional parallel resonant converters, *IEE Electronics Letters*, 24, 1510, 1988.

Luo, F.L. and Ye, H., Investigation of π-CLL current source resonant inverter, in *Proceedings of IEEE-IPEMC'03*, Xi'an China, 2003, p. 658.

Matsuo, M., Suetsugu, T., Mori, S., and Sasase, I., Class DE current-source parallel resonant inverter, *IEEE Transactions on IE*, 46, 242, 1999.

Severns, R.P., Topologies for three-element converters, *IEEE Transactions on PE*, 7, 89, 1992.

Van Wyk, J.D. and Snyman, D.B., High frequency link systems for specialized power control applications, in *Proceedings of IEEE-IAS'82 Annual Meeting*, 1982, p. 793.

4

Cascade Double Γ-CL Current Source Resonant Inverter

4.1 Introduction

This chapter introduces a four-element current source resonant inverter (CSRI): a cascade double Γ-type C-L circuit CSRI. Its circuit diagram is shown in Figure 4.1. It consists of four energy-storage elements, the double Γ-CL: C_1-L_1 and C_2-L_2. The energy source is a DC voltage V_{in} chopped by two main switches S_1 and S_2 to construct a bipolar current source, $i_i = \pm I_i$. The pump inductors L_{10} and L_{20} are equal to each other, and are large enough to keep the source current nearly constant during operation. The real load absorbs the delivered energy, its equivalent load should be proposed resistive, R_{eq}. The equivalent circuit diagram is shown in Figure 4.2.

4.2 Mathematic Analysis

In order to concentrate the function analysis of this double cascade Γ-CL CSRI, assume:

1. The inverter's source is a constant current source determined by the pump circuits.
2. Two MOSFETs in the switching circuit are turned-on and turned-off 180° out of phase with each other at same switching frequency and with a duty cycle of 50%.
3. Two switches are ideal components without on-resistance, and negligible parasitic capacitance and zero switching time.
4. Two diodes are components having a zero forward voltage drop and forward resistance.
5. Four energy-storage elements are passive, linear, time-invariant, and do not have parasitic reactive components.

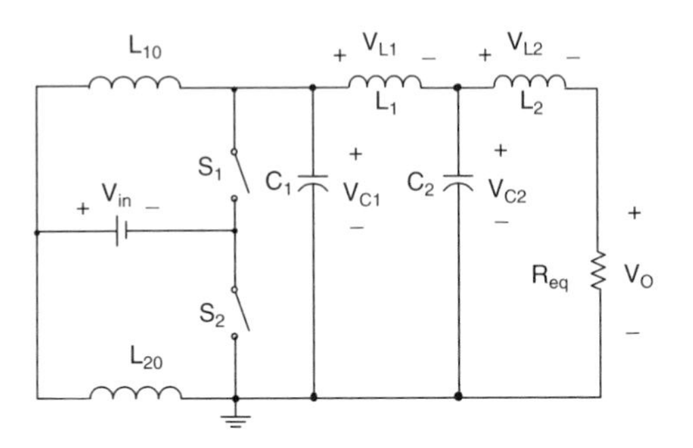

FIGURE 4.1
Cascade double Γ-CL CSRI.

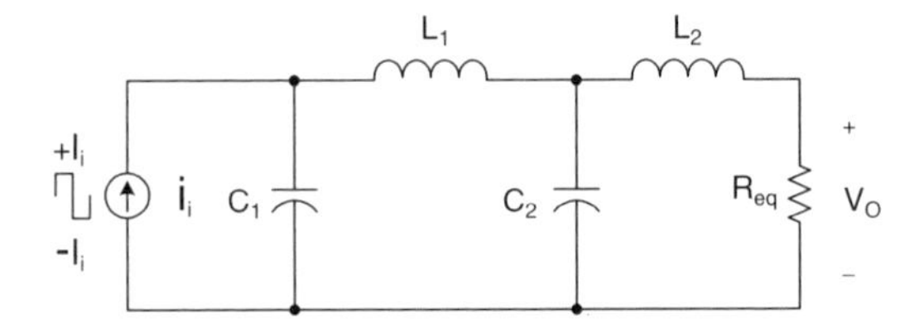

FIGURE 4.2
Equivalent circuit.

Based on these assumptions and the equivalent circuit the following analysis is derived.

4.2.1 Input Impedance

This CSRI is a fourth-order system. The mathematic analysis of operation and stability is more complex than three energy-storage element current source resonant inverters. The input impedance is given by

$$Z(\omega) = \frac{R_{eq}(1-\omega^2 L_1 C_2) + j\omega(L_1 + L_2 - \omega^2 L_1 L_2 C_2)}{\begin{pmatrix} 1-\omega^2(L_1C_1 + L_2C_1 + L_2C_2) + \omega^4 L_1 L_2 C_1 C_2 \\ + j\omega R_{eq}(C_1 + C_2 - \omega^2 L_1 C_1 C_2) \end{pmatrix}} \tag{4.1}$$

or

$$Z(\omega) = \frac{R_{eq}(1-\omega^2 L_1 C_2) + j\omega(L_1 + L_2 - \omega^2 L_1 L_2 C_2)}{B(\omega)} \tag{4.2}$$

where

$$B(\omega) = 1 - \omega^2(L_1C_1 + L_2C_1 + L_2C_2) + \omega^4 L_1L_2C_1C_2 + j\omega R_{eq}(C_1 + C_2 - \omega^2 L_1C_1C_2) \tag{4.3}$$

4.2.2 Components, Voltages, and Currents

This CSRI has four resonant components C_1, C_2, L_1 and L_2, plus the output equivalent resistance R_{eq}. In order to compare with the parameters easily, all components voltages and currents are responded to the input fundamental current I_i.

Voltage and current on capacitor C_1 is

$$\frac{V_{C1}(\omega)}{I_i(\omega)} = \frac{R_{eq}(1 - \omega^2 L_1C_2) + j\omega(L_1 + L_2 - \omega^2 L_1L_2C_2)}{B(\omega)} \tag{4.4}$$

$$\frac{I_{C1}(\omega)}{I_i(\omega)} = \frac{R_{eq}(1 - \omega^2 L_1C_2) + j\omega(L_1 + L_2 - \omega^2 L_1L_2C_2)}{B(\omega) / j\omega C_1} \tag{4.5}$$

Voltage and current on inductor L_1 is

$$\frac{V_{L1}(\omega)}{I_i(\omega)} = \frac{-R_{eq}\omega^2 L_1C_2 + j\omega L_1(1 - \omega^2 L_2C_2)}{B(\omega)} \tag{4.6}$$

$$\frac{I_{L1}(\omega)}{I_i(\omega)} = \frac{(1 - \omega^2 L_2C_2) + jR_{eq}\omega C_2}{B(\omega)} \tag{4.7}$$

Voltage and current on capacitor C_2 is

$$\frac{V_{C2}(\omega)}{I_i(\omega)} = \frac{R_{eq} + j\omega L_2}{B(\omega)} \tag{4.8}$$

$$\frac{I_{C2}(\omega)}{I_i(\omega)} = \frac{-\omega^2 L_2C_2 + jR_{eq}\omega C_2}{B(\omega)} \tag{4.9}$$

Voltage and current on inductor L_2 is

$$\frac{V_{L2}(\omega)}{I_i(\omega)} = \frac{j\omega L_2}{B(\omega)} \tag{4.10}$$

$$\frac{I_{L2}(\omega)}{I_i(\omega)} = \frac{1}{B(\omega)} \tag{4.11}$$

The output voltage and current on the resistor R_{eq}:

$$\frac{V_o(\omega)}{I_i(\omega)} = \frac{R_{eq}}{B(\omega)} \tag{4.12}$$

The current transfer gain is given by

$$g(\omega) = \frac{I_o(\omega)}{I_i(\omega)} = \frac{1}{B(\omega)} \tag{4.13}$$

4.2.3 Simplified Impedance and Current Gain

Usually, the input impedance and output current gain are paid more attention rather than other transfer functions listed in the previous section. To simplify the operation, select:

$$L_1 = L_2 = L; \quad C_1 = C_2 = C \quad \omega_0 = \frac{1}{\sqrt{LC}} \; Z_0 = \sqrt{\frac{L}{C}}$$

$$Q = \frac{Z_0}{R} = \frac{\omega_0 L}{R_{eq}} = \frac{1}{\omega_0 C R_{eq}} \qquad \beta = \frac{\omega}{\omega_0}$$

Obtain

$$B(\beta) = 1 - 3\beta^2 + \beta^4 + j\frac{2-\beta^2}{Q}\beta \tag{4.14}$$

Therefore,

$$Z = \frac{(1-\beta^2) + jQ(2-\beta^2)}{1 - 3\beta^2 + \beta^4 + j\frac{2-\beta^2}{Q}\beta} R_{eq} = |Z| < \phi \tag{4.15}$$

where

$$|Z| = \frac{\sqrt{(1-\beta^2)^2 + Q^2(2-\beta^2)^2}}{\sqrt{(1 - 3\beta^2 + \beta^4)^2 + \beta^2(\frac{2-\beta^2}{Q})^2}} R_{eq}$$

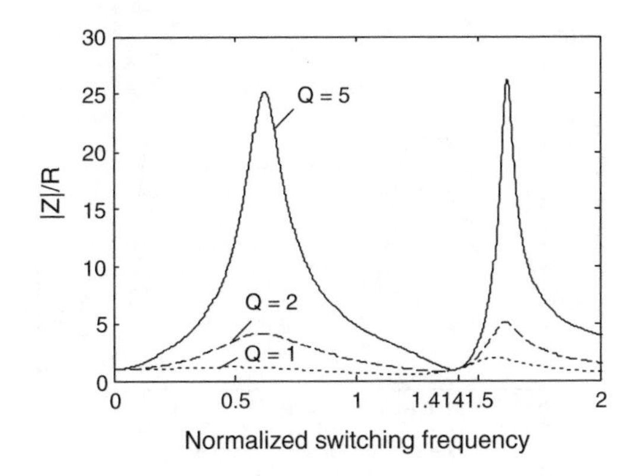

FIGURE 4.3
The curves of $|Z|/R_{eq}$ vs. β referring to Q.

TABLE 4.1
$|Z|/R_{eq}$ vs. β Referring to Various Q

	β = 0.59	**1**	**1.2**	**1.414**	**1.59**
$Q = 1$	1.1994	0.7071	0.5672	1.0000	2.0237
2	4.1662	1.7889	1.0955	1.0000	4.9137
5	23.464	4.9029	2.7031	1.0000	17.463

and
$$\phi = \tan^{-1}\frac{2-\beta^2}{1-\beta^2}Q - \tan^{-1}\frac{(2-\beta^2)\beta}{(1-3\beta^2+\beta^4)Q}$$

The characteristics of input impedance $|Z|/R_{eq}$ vs. relevant frequency β referring to various Q, is shown in Figure 4.3 and Table 4.1. The characteristics of phase angle ϕ vs. relevant frequency β referring to various Q, is shown in Figure 4.4 and Table 4.2. The current transfer gain becomes

$$g(\beta) = \frac{1}{1-3\beta^2+\beta^4+j\dfrac{2-\beta^2}{Q}\beta} = |g| < \theta \tag{4.20}$$

where

$$|g| = \frac{1}{\sqrt{(1-3\beta^2+\beta^4)^2+\beta^2(\dfrac{2-\beta^2}{Q})^2}} \tag{4.21}$$

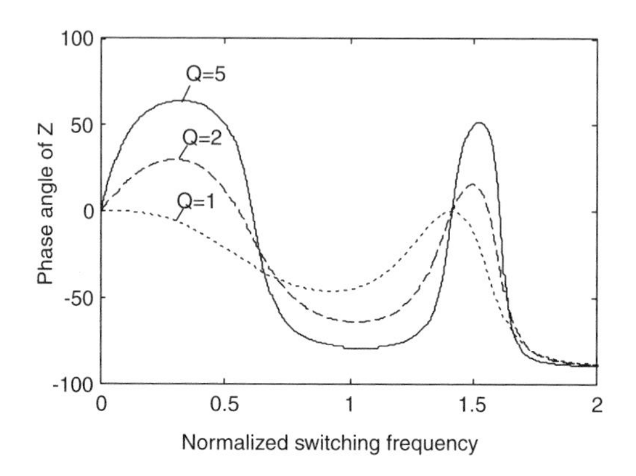

FIGURE 4.4
The curves of ϕ vs. β referring to Q.

TABLE 4.2
ϕ vs. β Referring to Various Q

	β = 0.59	1	1.2	1.414	1.59
Q = 1	–29.3	–45.0	–28.5	0.0	–48.3
2	–9.5	–63.4	–56.8	0.0	–17.6
5	13.9	–78.7	–76.4	0.0	29.0

and

$$\theta = -\tan^{-1}\frac{(2-\beta^2)\beta}{(1-3\beta^2+\beta^4)Q}$$

The characteristics of current transfer gain $|g|$ vs. relevant frequency β referring to various Q, is shown in Figure 4.5 and Table 4.3.

The characteristics of the phase angle θ vs. relevant frequency β referring to various Q, is shown in Figure 4.6 and Table 4.4.

For various β and Q, different current transfer gain is obtained. Actually, the maximum $|g|$ can be found from

$$\frac{d}{d\beta^2}|g| = 0$$

or

$$4\beta^6 + (\frac{3}{Q^2} - 18)\beta^4 + (22 - \frac{8}{Q^2})\beta^2 + (\frac{4}{Q^2} - 6) = 0 \tag{4.22}$$

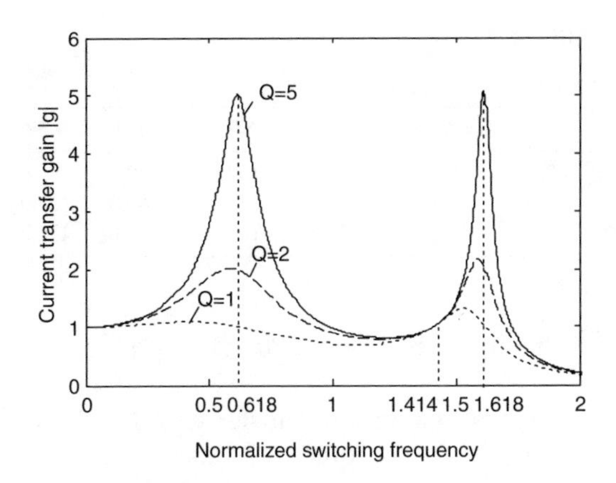

FIGURE 4.5
The curves of $|g|$ vs. β referring to Q.

TABLE 4.3
$|g|$ vs. β Referring to Various Q

	β = 0.59	1	1.2	1.414	1.59
Q = 1	1.0229	0.7071	0.7062	0.9994	1.1607
2	2.0270	0.8944	0.7747	0.9994	2.1641
5	4.7725	0.9806	0.7977	0.9994	3.9087

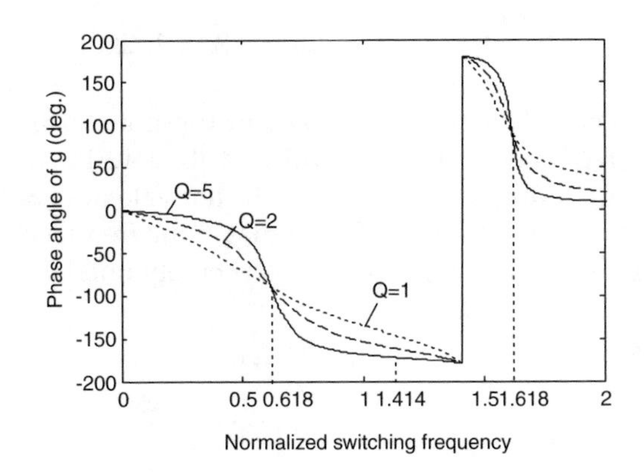

FIGURE 4.6
The curves of θ vs. β referring to Q.

when Q = 2:

$$4\beta^6 - 17.25\beta^4 + 20\beta^2 - 5 = 0 \tag{4.23}$$

TABLE 4.4

θ vs. β Referring to Various Q

	β = 0.59	1	1.2	1.414	1.59
Q = 0.5	−87.7	−116.6	−132.8	−180.0	96.6
1	−85.5	−135.0	−151.7	−180.0	102.9
2	−81.0	−153.4	−164.9	−180.0	114.7

yields

$$\beta_1 = 0.59 \quad \beta_2 = 1.20 \quad \beta_3 = 1.59$$

Take $\beta_3 = 1.59$ (the local maximum transfer gain is achieved at $\beta_1 = 0.59$), and the corresponding $|g(\beta)|$ is equal to 2.165.

In practice, Equation (4.23) is dependent on the qualify factor Q. For various load conditions, the maximum transfer gain will be achieved at different operating frequencies. When $Q >> 1$, we then have

$$2\beta^6 - 9\beta^4 + 11\beta^2 - 3 = 0 \tag{4.24}$$

or

$$(2\beta^2 - 3)(1 - 3\beta^2 + \beta^4) = 0$$

yields three positive real roots as

$$\beta_4 = 0.618 \quad \beta_5 = 1.618 \quad \beta_6 = 1.225$$

It should be noted that two peaks exist in the transfer gain curves with corresponding frequencies at $\beta_4 = 0.618$ and $\beta_5 = 1.618$, respectively. The current transfer gain drops from the peak to the vale at $\beta_6 = 1.225$.

Taking further investigation, it is found that at the frequencies corresponding to peak gain, the following equation can be obtained,

$$1 - 3\beta^2 + \beta^4 = 0 \tag{4.25}$$

Thus, the current transfer gain at β_4 and β_5 is

$$|g(\beta)| = \left. \frac{1}{\sqrt{(1 - 3\beta^2 + \beta^4)^2 + \left(\dfrac{2 - \beta^2}{Q}\right)^2 \beta^2}} \right|_{\beta_4, \beta_5} = Q \tag{4.26}$$

The relevant phase angle θ is

$$\theta_4 = -90^\circ \quad \theta_5 = 90^\circ$$

The results indicate that the current transfer gain is proportional to the quality factor Q. The larger the value of Q is, the higher the gain $|g(\beta)|$. For instance, when $Q = 1, 2$ and 5 with $\beta_4 = 0.618$ or $\beta_5 = 1.618$, $|g(\beta)|$ will have the same value. Note that although β_4 and β_5 are derived from the assumption of $Q >> 1$, the Equation (4.26) is still valid for all the values of quality factors.

Furthermore, when $Q << 1$, Equation (4.22) can be rearranged as

$$3\beta^4 - 8\beta^2 + 4 = 0 \tag{4.27}$$

giving other two positive real roots as

$$\beta_7 = 0.816 \quad \beta_8 = 1.414$$

As seen from Figure 4.5, the minimum transfer gain is achieved at β_7 while the maximum gain is obtained at β_8. These characteristic points will be useful in the estimation of the transfer gain curves. Notice that for low Q values, the frequency characteristics of Cascade Double Γ-CL CSRI approach those of conventional series-loaded resonant inverters, especially when β is near to 1.414.

When $\beta = \beta_c = 1.414$ $(\beta_c^2 = 2)$, all the curves will intersect at one point where the corresponding current transfer gain is

$$|g(\beta_c)| = \frac{1}{\sqrt{(1 - 3\beta_c^{\,2} + \beta_c^{\,4})^2 + \left(\dfrac{2 - \beta_c^{\,2}}{Q}\right)^2 \beta_c^{\,2}}} \equiv 1 \tag{4.28}$$

This point is always called load-independent point since the current transfer gain keeps constant with any value of quality factor Q.

If the inverter is working at the conditions:

$$\beta = 1 \text{ and } Q >> 1$$

we then have

$$Z = \frac{jQ}{-1 + j/Q} R_{eq} \approx -jQR_{eq}$$
$$\varphi = \frac{\pi}{2} - \tan^{-1}\frac{1}{-Q} \approx \frac{\pi}{2} - \pi \approx -\frac{\pi}{2} \tag{4.29}$$

and correspondingly

$$g(\beta) = \frac{1}{-1 + j/Q} = -1$$
$$\theta = -\pi \tag{4.30}$$

The double Γ-CL circuit is not only the resonant circuit, but the band-pass filter as well. All higher order harmonic components in the input current are effectively filtered by the double Γ-CL circuit. The output current is nearly a pure sinusoidal waveform with the fundamental frequency $\omega = 2f$.

4.2.4 Power Transfer Efficiency

The power transfer efficiency is a very important parameter and it is calculated here. From Figure 4.2, the input current is a bipolar value $i_i(\omega t)$:

$$i_i(\omega t) = \begin{cases} I_i & 2n\pi \le \omega t \le (2n+1)\pi \\ -I_i & (2n+1)\pi \le \omega t \le 2(n+1)\pi \end{cases} \quad \text{with } n = 0, 1, 2, 3, \ldots \infty \tag{4.31}$$

where $I_i = V_1 / (Z + j\omega L_{10})$ ($L_{10} = L_{20}$) that varies with operating frequency.

This is a square waveform pulse-train. Applying fast Fourier transform (FFT), the spectrum form is

$$i_i(\omega t) = \frac{4I_i}{\pi} \sum_{n=0}^{\infty} \frac{\sin(2n+1)\omega t}{2n+1} \tag{4.32}$$

The fundamental frequency component is

$$i_{fund}(\omega t) = \frac{4I_i}{\pi} \sin \omega t \tag{4.33}$$

Output current is

$$i_0(\omega t) = g \frac{4I_i}{\pi} \sin \omega t \tag{4.34}$$

The power transfer efficiency from input current source to AC output load is analyzed and calculated. Since the input current is a square waveform the total input power is

$$P_{in} = I_i^2 |Z| \tag{4.35}$$

The output current is nearly a pure sinusoidal waveform, its root-mean-square value is its peak value times $1/\sqrt{2}$. Therefore the output power is

$$P_O = (|g|\frac{4I_i}{\pi\sqrt{2}})^2 R_{eq} = 8(|g|\frac{I_i}{\pi})^2 R_{eq} \tag{4.36}$$

We can get the power transfer efficiency as

$$\eta = \frac{P_O}{P_{in}} = \frac{8\frac{(I_i|g|)^2}{\pi^2}R_{eq}}{I_i^2|Z|} = \frac{8|g|^2 R_{eq}}{\pi^2|Z|} \tag{4.37}$$

Considering Equation (4.15) and Equation (4.21), we obtain:

$$\eta = \frac{8}{\pi^2}\frac{1}{\sqrt{(1-\beta^2)^2 + Q^2\beta^2(2-\beta^2)^2}\sqrt{(1-3\beta^2+\beta^4)^2 + \beta^2(\frac{2-\beta^2}{Q})^2}} \tag{4.38}$$

If $\beta^2 = 2$ ($\beta = 1.414$) with any Q, $\eta = 0.8106$.
If $\beta^2 = 1$ ($\beta = 1$) and $Q = 1$, $\eta = 0.5732$.
If $\beta^2 = 2.5$ ($\beta = 1.581$) and $Q = 1$, $\eta = 0.5771$.
If $\beta^2 = 2.618$ ($\beta = 1.618$) and $Q = 1$, $\eta = 0.4263$.

Therefore, the characteristics of efficiency η vs. relevant frequency β referring to various Q is obtained, and shown in Figure 4.7 and Table 4.5.

4.3 Simulation Result

In order to verify analysis and calculation, using PSpice software simulation method to obtain a set of simulation waveforms as shown in Figure 4.8 to Figure 4.9 corresponding to $Q = 2$ and $\beta = 1$, 1.414, and 1.59. The parameter values are set below:

$$I = 1\text{ A},\ V_1 = 30\text{ V},\ L_{10} = L_{20} = 20\text{ mH},\ R_{eq} = 10\ \Omega,$$
$$C_1 = C_2 = C = 0.22\ \mu\text{F, and } L_1 = L_2 = L = 100\ \mu\text{H}.$$

Therefore, $\omega_0 = 213$ krad/s, $f_0 = 33.93$ kHz, and $Q = 2$. The particular frequencies for the figures are $f = 33.9$ kHz, 48.0 kHz and 54.0 kHz. In order to pick the input current a small resistance $R_0 = 0.001\ \Omega$ is employed. The load in the simulation circuit is R rather than R_{eq}.

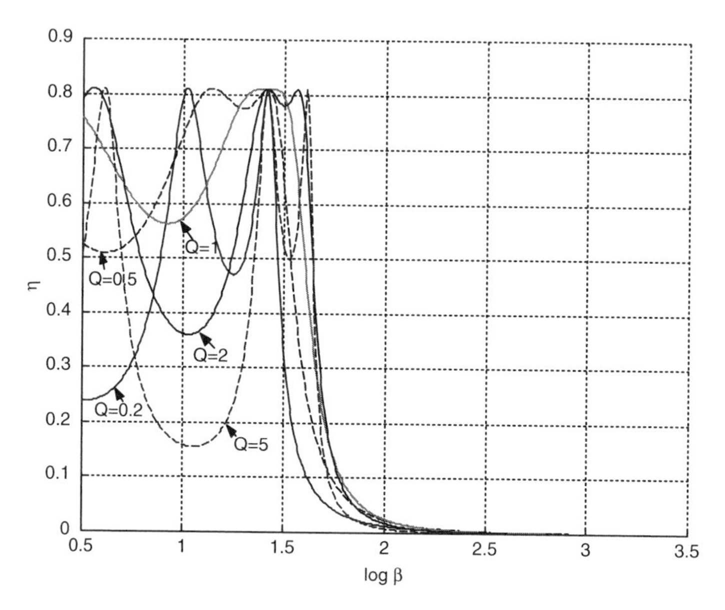

FIGURE 4.7
Curves of η vs. β referring to *Q*.

TABLE 4.5
η vs. β Referring to Various *Q*

	β = 0.618	**0.8**	**1**	**1.414**	**1.581**	**1.618**	**2**	**3**	**4**	**5**
Q = 0.2	0.2496	0.3527	0.7948	0.8106	0.1359	0.0995	0.0127	0.0008	0.0001	0.0000
0.5	0.5098	0.5559	0.7250	0.8106	0.3268	0.2394	0.0238	0.0009	0.0001	0.0000
1	0.6895	0.5885	0.5732	0.8106	0.5771	0.4263	0.0253	0.0006	0.0001	0.0000
2	0.7745	0.4927	0.3625	0.8106	0.7955	0.6304	0.0176	0.0003	0.0000	0.0000
5	0.8045	0.2680	0.1590	0.8106	0.6473	0.7715	0.0079	0.0001	0.0000	0.0000

All figures have two parts a and b. Figure a shows the input and output current waveforms, and b shows the corresponding FFT spectrum. The first channel of Figure 4.11a to Figure 4.13a is the input current flowing through resistance R_0, which corresponds to the input current $i_i(\omega t)$. It is a square waveform pulse train with the pulse-width $\omega t = \pi$. The second channel of Figure 4.11a to Figure 4.13a is the output current flowing through resistance *R*. From Figure 4.11b to Figure 13b, the first channel of each figure is the corresponding input current FFT spectrum. The second channel of each figure is the corresponding output current FFT spectrum. From the spectrums, it can be clearly seen that there is only mono-frequency existing in output currents. All output current waveforms are very pure sinusoidal function.

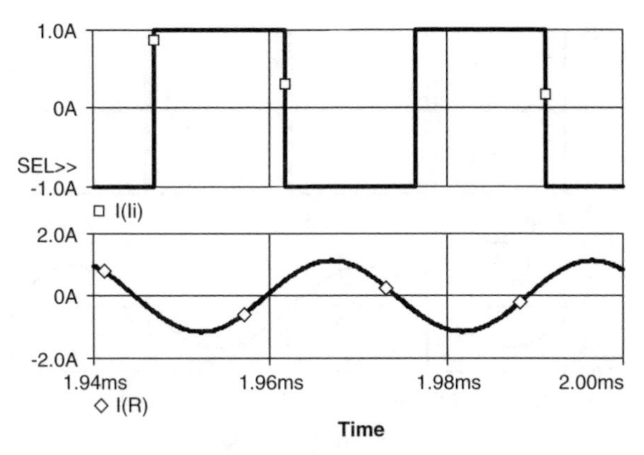

(a) Input and output current waveforms

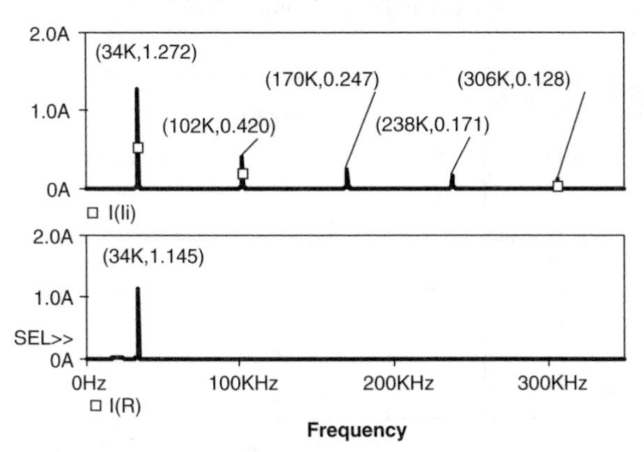

(b) The corresponding FFT spectrum of input and output current waveforms

FIGURE 4.8
Input and output current waveforms at f = 33.9 kHz.

4.3.1 β = 1, *f* = 33.9 kHz, *T* = 29.5 μs

The waveforms and corresponding FFT spectrums are shown in Figure 4.8. The THD = 0 and current transfer gain is

$$g = \frac{I(R)}{I(R0)_1}\Big|_{f=33.9kHz} = \frac{1.145}{1.272} = 0.9002$$

4.3.2 β = 1.4142, *f* = 48.0 kHz, *T* = 20.83 μs

The waveforms and corresponding FFT spectrums are shown in Figure 4.9. The THD = 0 and current transfer gain is

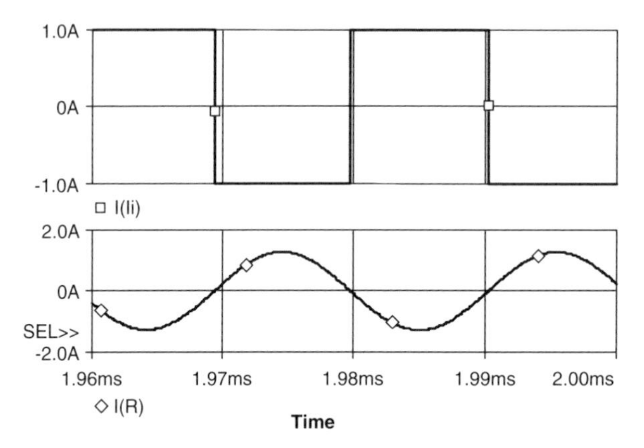

(a) Input and output current waveforms

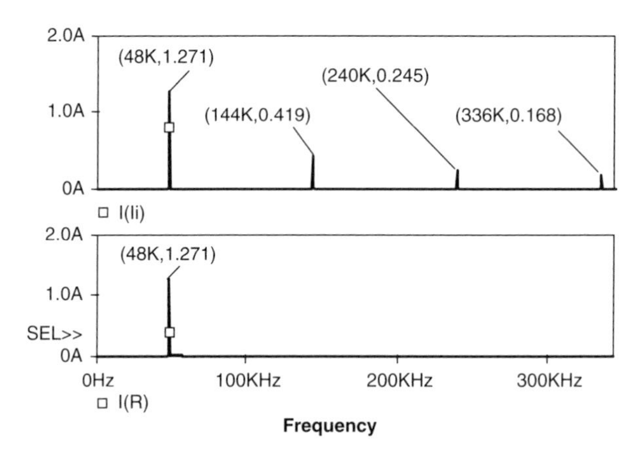

(b) The corresponding FFT spectrum of input and output current waveforms

FIGURE 4.9
Input and output current waveforms at f = 48 kHz.

$$g = \frac{I(R)}{I(R0)_1}\Big|_{f=48kHz} = \frac{1.271}{1.271} = 1.00$$

4.3.3 β = 1.59, f = 54 kHz, T = 18.52 μs

The waveforms and corresponding FFT spectrums are shown in Figure 4.10. The THD = 0 and current transfer gain is

$$g = \frac{I(R)}{I(R0)_1}\Big|_{f=54kHz} = \frac{2.752}{1.273} = 2.162$$

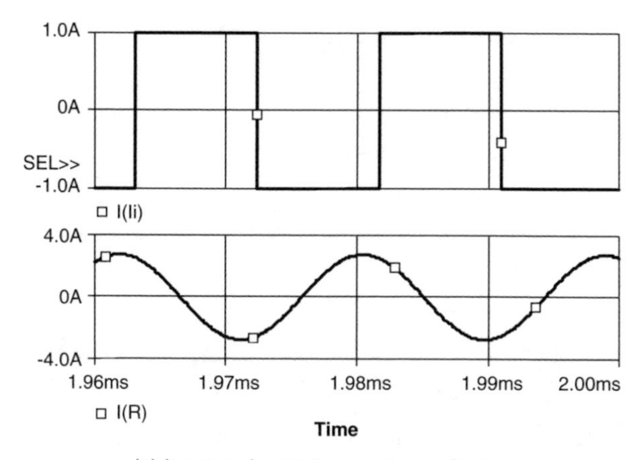

(a) Input and output current waveforms

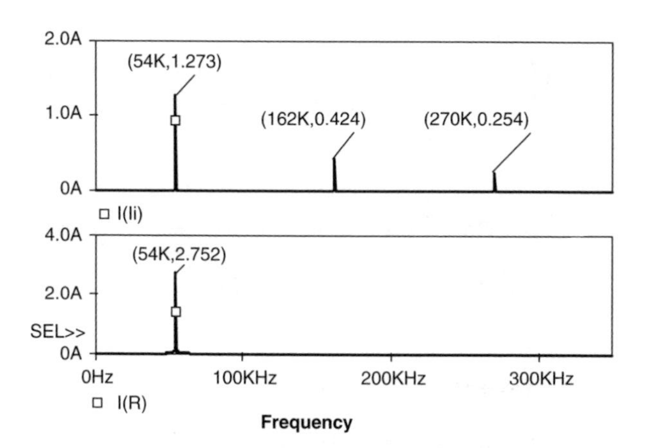

(b) The corresponding FFT spectrum of input and output current waveforms

FIGURE 4.10
Input and output current waveforms at $f = 54$ kHz.

4.4 Experimental Result

In order to verify our analysis and calculation, a test rig with the same components was constructed:

$$I = 1\text{ A},\ V_1 = 30\text{ V},\ L_{10} = L_{20} = 20\text{ mH},\ R_{eq} = 10\ \Omega,$$
$$C_1 = C_2 = C = 0.22\ \mu\text{F, and } L_1 = L_2 = L = 100\ \mu\text{H}.$$

Therefore, $\omega_0 = 213$ krad/s, $f_0 = 33.93$ kHz, and $Q = 2$. The MOSFET device is IRF640 ($R_{ds} = 0.15$ ohm, $C_{ds} = 140$ pF). Since the junction capacitance of C_{ds}

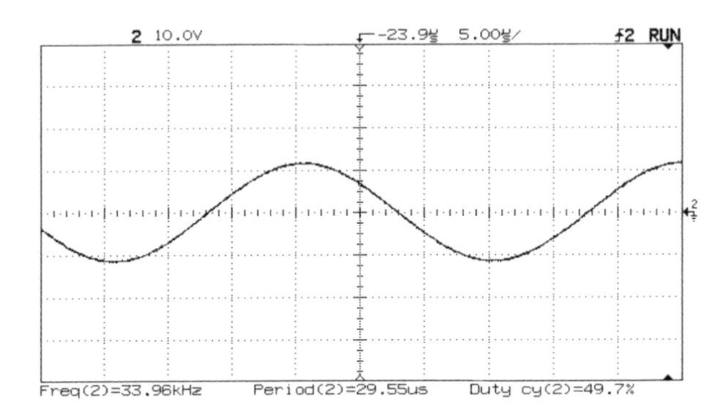

FIGURE 4.11
Testing waveform of the output voltage at $\beta = 1$.

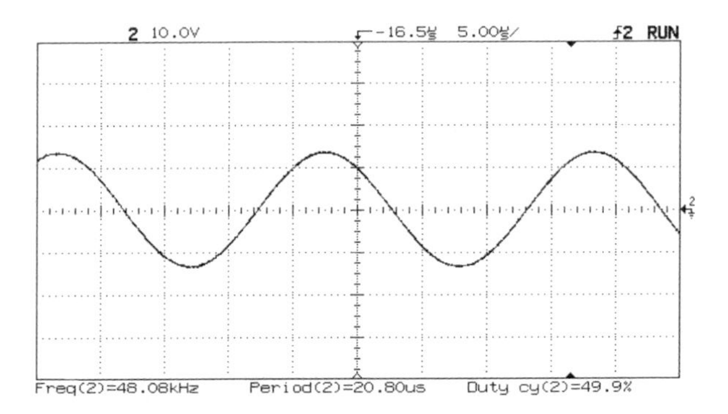

FIGURE 4.12
Testing waveform of the output voltage at $\beta = 1.414$.

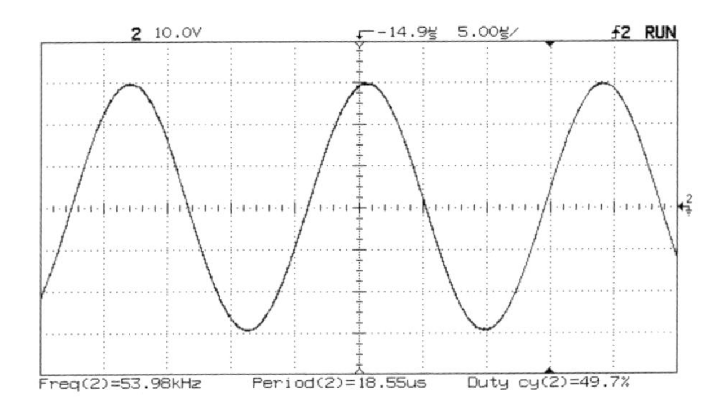

FIGURE 4.13
Testing waveform of the output voltage at $\beta = 1.59$.

is much smaller than the resonance capacitance $C = 0.22\ \mu F$ ($C/C_{ds} = 1429$), it does not affect the experimental results. The output current waveform is a perfect sine function. A set of tested output voltage waveforms that correspond to the output current with $\beta = 1$, 1.4142, and 1.59 is shown in Figure 4.11 to Figure 4.13. The particular applied frequencies for the figures are $f = 33.9$ kHz, 48.0 kHz, and 54.0 kHz, and peak-to-peak voltage V_{pp} = 28.1 V, 40.3 V, 71.9 V, and 65.5 V, which are very close to the values in Figure 4.8 to Figure 4.10.

4.5 Discussion

4.5.1 Function of the Double Γ-CL Circuit

Single Γ-CL filter is a well-known circuit. As a Π-CLC filter, it is a typical low-pass filter. All harmonics with frequency $\omega > \omega_0$ will be blocked. Cascade double Γ-CL filter circuit has thoroughly different characteristics from that of low-pass filters. It allows the signal with higher frequency $\omega > \omega_0$ (it means $\beta > 1$ in Section 4.2) passing it and enlarging the energy.

4.5.2 Applying Frequency to This Double Γ-CL CSRI

From analysis and verifications it can be found the fact that the effective applying frequency to this double Γ-CL current source resonant inverter is (0.6 to 2.0) f_0. Outside this region both current transfer gain and efficiency is falling fast.

4.5.3 Explanation of $g > 1$

We recognized the fact that current transfer gain is greater than unity from mathematical analysis, and simulation and experimental results. The reason to enlarge the fundamental current is that the resonant circuit transfers the energy of other higher order harmonics to the fundamental component. Therefore the gain of the fundamental current can be greater than unity.

Bibliography

Bhat, A.K.S., Analysis and design of a series-parallel resonant converter, *IEEE Transactions on Power Electronics*, 8, 1, 1993.

Bhat, A.K.S. and Swamy, M.M., Analysis and design of a high-frequency parallel resonant converter operating above resonance, in *Proceedings of IEEE Applied Power Electronics Conference*, 1988, p. 182.

Forsyth, A.J. and Ho, Y.K. E., Dynamic characteristics and closed-loop performance of the series-parallel resonant converter, *IEE Proceedings-Electric Power Applications*, 143, 345, 1996.

Ho, W.C. and Pong, M.H., Design and analysis of discontinuous mode series resonant converter, in *Proceedings of the IEEE International Conference on Industrial Technology*, 1994, p. 486.

Hua, G., Yang, E.X., Jiang, Y., and Lee, F.C.Y., Novel zero-voltage-transition PWM converters, *IEEE Transaction on Power Electronics*, 9, 213, 1994.

Keown, J., *OrCAD PSpice and Circuit Analysis*, 4th ed., New Jersey, Prentice Hall, 2001.

Liu, K.H., Oruganti, R., and Lee, F.C., Resonant switches-topologies and characteristics, in *Proceedings of IEEE Power Electronics Specialists Conference*, 1985, p. 106.

Luo F.L. and Wan, J.Z., Bipolar current source applying to cascade double gamma-CL CSRI, in *Proceedings of UROP Congress'2003*, Singapore, 2003, p. 328.

Luo, F.L. and Ye, H., Analysis of a double Γ-CL current source resonant inverter, in *Proceedings of IEEE IAS Annual Meeting IAS-2001*, Chicago, U.S., 2001, p. 289.

Luo, F.L. and Ye, H., Investigation and verification of a double Γ-CL current source resonant inverter, *Proceedings of IEE on Electric Power Applications*, 149, 369, 2002.

Luo, F.L. and Zhu, J.H., Verification of a double Γ-CL current source resonant inverter, in *Proceedings of IEE-IPEC'2003*, Singapore, 2003, p. 386.

Oruganti, R. and Lee, F.C., State-plane analysis of parallel resonant converter, in *Proceedings of IEEE Power Electronics Specialists Conference*, 1985, p. 56.

Vorpérian, V., *Analysis of Resonant Converter*, Ph.D dissertation, California Institute of Technology, Pasadena, May 1984.

5

Cascade Reverse Double Γ-LC Resonant Power Converter

A four-element power resonant converter — cascade reverse double Γ-LC resonant power converter (RPC) — will be discussed in this chapter. Since the first element is an inductor, the power supply should be a bipolar voltage source. Do remember that the first element of CSRIs in previous chapters is a capacitor, therefore, a bipolar current source was employed. The major work is concentrated on the analysis of steady-state operation, dynamic behavior, and control specialties of the novel resonant converter. The simulation and experimental results show that this resonant converter has many distinct advantages over the existing two or three element resonant converters and overcomes their drawbacks.

5.1 Introduction

Generally, a switched-mode power converter is often required to meet all or most of the following specifications:

- High switching frequency
- High power density for reduction of size and weight
- High conversion efficiency
- Low total harmonic distortion (THD)
- Controlled power factor if the source is an AC voltage
- Low electromagnetic interference (EMI)

A review of the commonly used pulse-width-modulating (PWM) converter and new generated resonant converter is presented in order to fully understand the two major branches of the high frequency switching converter. Although PWM technique is widely used in power electronic applications, it encounters serious problems when the switching converter operates at

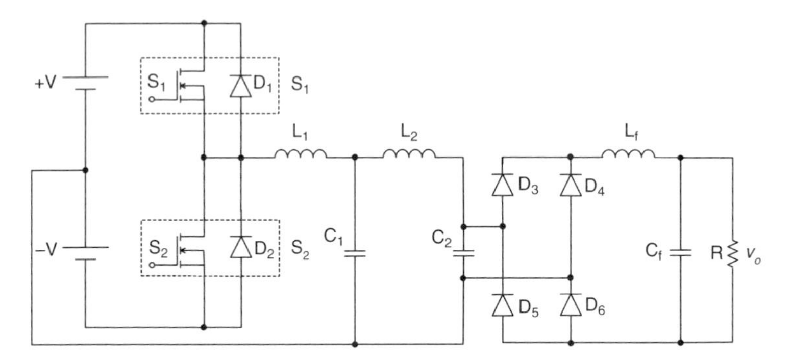

FIGURE 5.1
Circuit diagram of cascade reverse double Γ-LC RPC.

high frequencies. Due to the hard-switching transitions caused by PWM technique, switching losses possess large proportion in total power dissipations. In other words, when the switch is turned on, the current through it rises very fast, while the voltage across it cannot descend immediately due to the parasitic output capacitance. Similarly, when the switch is turned off, the voltage across it rises rapidly while the current through it cannot drop at once because of the recombination of carriers.

In general, a resonant power converter is defined as a converter in which one or more switching waveforms are resonant waveforms. It is reasonable to say that a resonant power converter usually contains a resonant circuit. In fact, there are many topologies of the resonant power converter, which are many more than ZCS and ZVS resonant converters.

5.2 Steady-State Analysis of Cascade Reverse Double Γ-LC RPC

In this chapter, a cascade reverse double Γ-LC RPC is introduced. Under some assumptions and simplifications, the steady-state AC analysis is undertaken to study the two most interesting topics: voltage transfer gain and the input impedance.

5.2.1 Topology and Circuit Description

The circuit diagram of a cascade reverse double Γ-LC resonant power converter with a diode-bridge rectifier plus load is shown in Figure 5.1. Like other resonant converters, this topology consists of a bipolar voltage source, resonant network — the cascade reverse double Γ-LC, the rectifier-plus-filter, and load (such as a resistive load R). The power MOSFETs S_1 and S_2 and their antiparalleled diodes D_1 and D_2 act together as a bipolar voltage source. To operate this circuit, S_1 is turned on and off 180° out of phase with respect to the turn-off and -on of S_2 at same frequency $\omega = 2\pi f$. After the switching circuit, the input voltage can be considered as a bipolar square-wave voltage alternating in value between +V and –V, which is then input to the resonant circuit section. L_1, L_2 and C_1, C_2 represent the resonant inductors and capacitors, respectively. The output DC voltage is obtained by rectifying the voltage across the second resonant capacitor C_2. L_f and C_f comprise a low-pass filter to smooth out the output voltage and current, and R denotes either an actual or an equivalent load resistance.

The following assumptions should be made:

1. All the switches and diodes used in the converter are ideal components
2. All the inductors and capacitors are passive, linear and time-invariant
3. The output inductor is large enough to assume that the load current does not vary significantly during switching period
4. The converter operates above resonance

5.2.2 Classical Analysis on AC Side

This analysis is available on the AC side before the rectifier bridge.

5.2.2.1 Basic Operating Principles

For the cascade reverse double Γ-LC RPC considered here, the half-bridge converter applies a square wave of voltage to a resonant network. Since the resonant network has the effect of filtering the higher-order harmonic voltages, a sine wave of current will appear at the input to the resonant circuit (this is true over most of the load range of interest). This fact allows classical AC analysis techniques to be used. The analysis proceeds as follows. The fundamental component of the square wave input voltage is applied to the resonant network, and the resulting sine waves of current and voltage in the resonant circuit are computed using classical AC analysis. For a rectifier with an inductor output filter, the sine wave voltage at the input to the rectifier is rectified, and the average value is taken to arrive at the resulting DC output voltage. For a capacitive output filter, a square wave of voltage appears at the input to the rectifier while a sine wave of current is injected into the rectifier. For this case the fundamental component of the square wave voltage is used in the AC analysis.

5.2.2.2 Equivalent Load Resistance

It is necessary that the rectifier with its filter should be expressed as an equivalent load resistance before the analysis is carried out, which illustrates

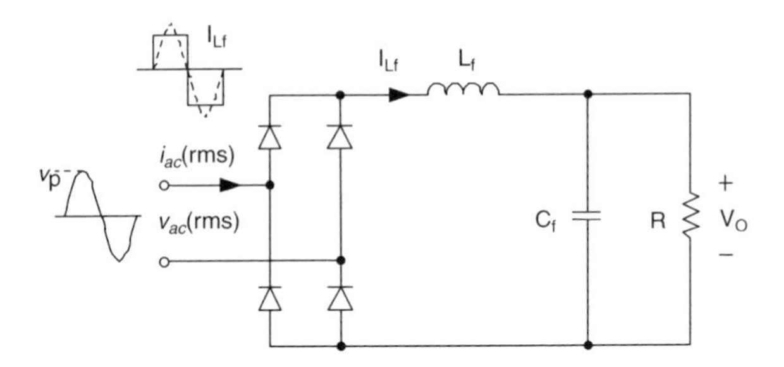

FIGURE 5.2
Equivalent load resistance R_{eq}.

the derivation of the equivalent resistance to use in loading the resonant circuit. The resonant converter uses an inductor output filter and drives the rectifier with an equivalent voltage source, i.e., a low-impedance source provided by the resonant capacitor. A square wave of current is drawn by the rectifier, and its fundamental component must be used in arriving at an equivalent AC resistance. For this case, the root-mean-square value of the voltage and current before the rectifier are given as:

$$v_{ac}(rms) = \frac{\pi}{2\sqrt{2}} V_o$$

$$i_{ac}(rms) = \frac{2\sqrt{2}}{\pi} I_{Lf}$$

$$R_{eq} = \frac{v_{ac}(rms)}{i_{ac}(rms)} = \frac{\pi^2}{8} \frac{V_o}{I_{Lf}} = \frac{\pi^2}{8} R$$

The equivalent resistance R_{eq} is shown in Figure 5.2.

5.2.2.3 Equivalent AC Circuit and Transfer Functions

The equivalent AC circuit diagram of the cascade reverse double Γ-LC RPC is shown in Figure 5.3. Note that all the parameters and variables are transferred to the s-domain. Using Laplace operator $s = j\omega$, it is a simple matter to write down the voltage transfer gain of the cascade reverse double Γ-LC RPC:

$$G(s) = \frac{V_{C2}(s)}{V_i(s)} = R_{eq} \Big/ [s^4 L_1 L_2 R_{eq} C_1 C_2 + s^3 L_1 L_2 C_1 + s^2 (L_1 R_{eq} C_1 + L_1 R_{eq} C_2 + L_2 R_{eq} C_2) + s(L_1 + L_2) + R_{eq}]$$

or

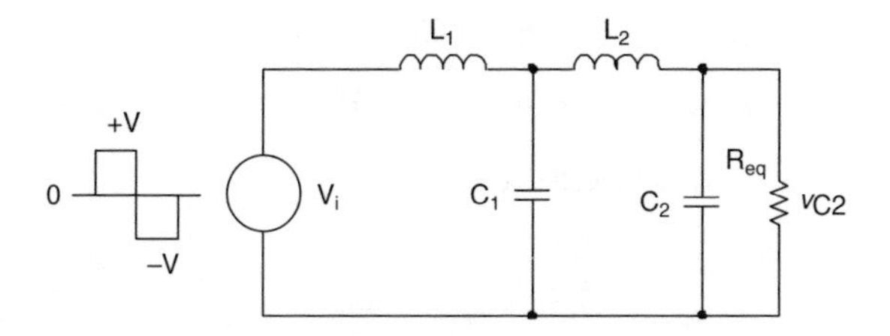

FIGURE 5.3
Equivalent circuit of the cascade reverse double Γ-LC RPC.

$$g(s) = R_{eq} / B(s) \tag{5.1}$$

where

$$\begin{aligned} B(s) = {} & s^4 L_1 L_2 R_{eq} C_1 C_2 + s^3 L_1 L_2 C_1 \\ & + s^2 (L_1 R_{eq} C_1 + L_1 R_{eq} C_2 + L_2 R_{eq} C_2) + s(L_1 + L_2) + R_{eq} \end{aligned} \tag{5.2}$$

The voltage and current stresses on different reactive resonant components are obtained with respect to the input fundamental voltage V_i.

Voltage and current on inductor L_1:

$$\frac{V_{L1}(s)}{V_i(s)} = \frac{sL_1[s^3 L_2 R_{eq} C_1 C_2 + s^2 L_2 C_1 + sR_{eq}(C_1 + C_2) + 1]}{B(s)} \tag{5.3}$$

$$\frac{I_{L1}(s)}{V_i(s)} = \frac{s^3 L_2 R_{eq} C_1 C_2 + s^2 L_2 C_1 + sR_{eq}(C_1 + C_2) + 1}{B(s)} \tag{5.4}$$

Voltage and current on capacitor C_1:

$$\frac{V_{C1}(s)}{V_i(s)} = \frac{s^2 L_2 R_{eq} C_2 + sL_2 + R_{eq}}{B(s)} \tag{5.5}$$

$$\frac{I_{C1}(s)}{V_i(s)} = \frac{sC_1(s^2 L_2 R_{eq} C_2 + sL_2 + R_{eq})}{B(s)} \tag{5.6}$$

Voltage and current on inductor L_2:

$$\frac{V_{L2}(s)}{V_i(s)} = \frac{sL_2(sR_{eq} C_2 + 1)}{B(s)} \tag{5.7}$$

$$\frac{I_{L2}(s)}{V_i(s)} = \frac{sR_{eq}C_2 + 1}{B(s)} \tag{5.8}$$

Voltage and current on capacitor C_2:

$$G(S) = \frac{V_{C2}(s)}{V_i(s)} = \frac{R_{eq}}{B(s)} \tag{5.9}$$

$$\frac{I_{C2}(s)}{V_i(s)} = \frac{R_{eq}}{B(s)C_2} \tag{5.10}$$

Voltage and current on the load R_{eq}:

$$G(s) = \frac{V_{Req}(s)}{V_i(s)} = \frac{R_{eq}}{B(s)}$$

$$\frac{I_{Req}(s)}{V_i(s)} = \frac{1}{B(s)}$$

The input impedance is given by

$$Z_{in} = \frac{B(s)}{s^3L_2R_{eq}C_1C_2 + s^2L_2C_1 + sR_{eq}(C_1 + C_2) + 1} \tag{5.11}$$

5.2.2.4 Analysis of Voltage Transfer Gain and the Input Impedance

In general, the voltage gain and the input impedance have more attractions to the designer rather than other transfer functions listed above. To simplify the mathematical analysis, the resonant components are chosen as follows:

$$L_1 = L_2 = L \qquad C_1 = C_2 = C \qquad \omega_0 = \sqrt{\frac{1}{LC}}$$

The quality factor Q is defined as

$$Q = \frac{\omega_0 L}{R_{eq}} = \frac{1}{\omega_0 C R_{eq}} = \frac{Z_0}{R_{eq}} \tag{5.12}$$

where the characteristic impedance Z_0 is

$$Z_0 = \sqrt{\frac{L}{C}} \tag{5.13}$$

The relative switching frequency is defined as

$$\beta = \frac{\omega}{\omega_0} \tag{5.14}$$

where ω is the switching frequency and ω_0 is the natural resonant radian frequency:

$$\omega_0 = \frac{1}{\sqrt{LC}} \tag{5.15}$$

Under the above-simplified conditions, the voltage gain $g(s)$ given in previous section can be rewritten as

$$B(\beta) = R_{eq}[1 - 3\beta^2 + \beta^4 + j(2 - \beta^2)\beta Q]$$

$$g(\beta) = \frac{1}{(1 - 3\beta^2 + \beta^4) + j(2 - \beta^2)\beta Q} = |g(\beta)| \angle \theta \tag{5.17}$$

The determinant and phase of $g(\beta)$ are given by:

$$|g(\beta)| = \frac{1}{\sqrt{(1 - 3\beta^2 + \beta^4)^2 + (2 - \beta^2)^2\beta^2 Q^2}}$$

and

$$\theta = -\tan^{-1}\frac{(2 - \beta^2)\beta Q}{1 - 3\beta^2 + \beta^4} \tag{5.18}$$

Analogiously, the input impedance $Z_{in}(s)$ can also be simplified as:

$$Z_{in}(\beta) = \frac{R_{eq}[1 - 3\beta^2 + \beta^4 + j(2 - \beta^2)\beta Q]}{1 - \beta^2 + j(2 - \beta^2)\beta / Q} = |Z_{in}(\beta)| \angle \phi \tag{5.19}$$

where

$$|Z_{in}(\beta)| = \frac{R_{eq}\sqrt{(1 - 3\beta^2 + \beta^4)^2 + (2 - \beta^2)^2\beta^2 Q^2}}{\sqrt{(1 - \beta^2)^2 + (2 - \beta^2)^2\beta^2 / Q^2}}$$

and

$$\phi = \tan^{-1}\frac{(2 - \beta^2)\beta Q}{1 - 3\beta^2 + \beta^4} - \tan^{-1}\frac{(2 - \beta^2)\beta}{(1 - \beta^2)Q} \tag{5.20}$$

The characteristics of the voltage gain $|g(\beta)|$ and phase angle θ vs. relative frequency β referring to various Q is shown in Figure 5.4 and Figure 5.5, respectively. Note that for lower Q value, the voltage gain $|g(\beta)|$ is higher than unity at some certain switching frequencies. It means the output voltage

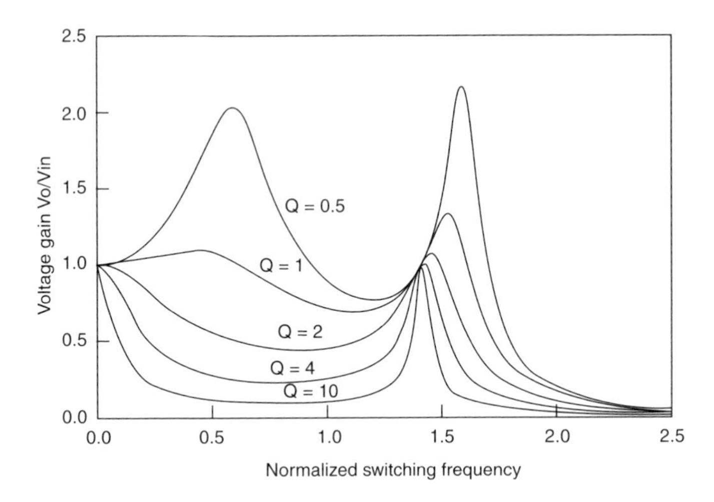

FIGURE 5.4
Voltage gain $|g(\beta)|$ vs. β referring to Q.

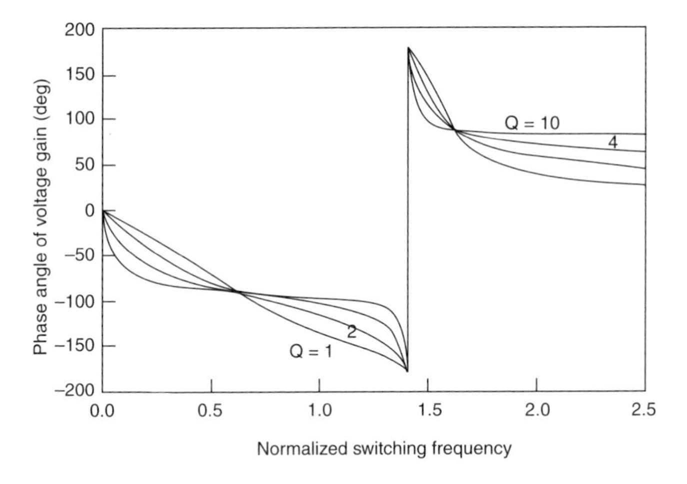

FIGURE 5.5
Phase angle θ vs. β referring to Q.

can be larger than the fundamental harmonic of the input voltage. The result could be explained thusly: the resonant network consisting of inductors and capacitors has the function of filtering the higher order harmonic components in the input quasi-square voltage. The energy of higher order harmonics is then transferred to the fundamental component thus enlarging the output voltage.

This explanation is reasonable when fast Fourier transform (FFT) is applied to the comparison to waveforms of the input voltage and the second capacitor voltage respectively. The resonant circuit in cascade reverse double Γ-LC

RPC allows the signal with higher frequency ($\beta > 1$) passing it and enlarging the energy. In other words, the bandwidth of this novel converter is wider, as it provides more options for the designer to choose the appropriate operating frequency in different applications.

The maximum voltage gain $|g(\beta)|$ can be obtained from

$$\frac{d}{d\beta^2}|g(\beta)| = 0$$

or

$$4\beta^6 + (3Q^2 - 18)\beta^4 + (22 - 8Q^2)\beta^2 + 4Q^2 - 6 = 0 \tag{5.21}$$

when

$$Q = 1 \quad 4\beta^6 - 15\beta^4 + 14\beta^2 - 2 = 0 \tag{5.22}$$

yields

$$\beta_1 = 0.42 \quad \beta_2 = 1.11 \quad \beta_3 = 1.53$$

It means the local maximum or minimum values on the gain curve are achieved at these roots, respectively. Generally, the voltage gain decreases with the increase of the Q value. For instance, when $Q = 1$, 2, and 4 with $\beta^2 = 2$, the gain $|g(\beta)| = 1$, 0.5, and 0.25, respectively. At certain switching frequencies, the smaller the value of Q is, the higher the gain.

Specifically, when $Q << 1$, Equation (5.21) can be simplified,

$$2\beta^6 - 9\beta^4 + 11\beta^2 - 3 = 0 \tag{5.23}$$

It yields two positive real roots as

$$\beta_1 = 0.618 \quad \beta_2 = 1.618$$

These roots indicate the existence of two peaks in the voltage gain curve, which is the main characteristics of the four-energy-storage-element resonant converters and rarely stated in most conventional two or three elements counterparts.

The voltage gain $|g(\beta)|$ at these two roots is then given by

$$|g(\beta)| = \left.\frac{1}{\sqrt{(1 - 3\beta^2 + \beta^4)^2 + (2 - \beta^2)^2\beta^2Q^2}}\right|_{\beta_1,\beta_2} = \frac{1}{Q} \tag{5.24}$$

This result is derived under the condition $Q << 1$, but applicable for all Q values.

Furthermore, when $Q >> 1$, Equation (5.21) can be rearranged,

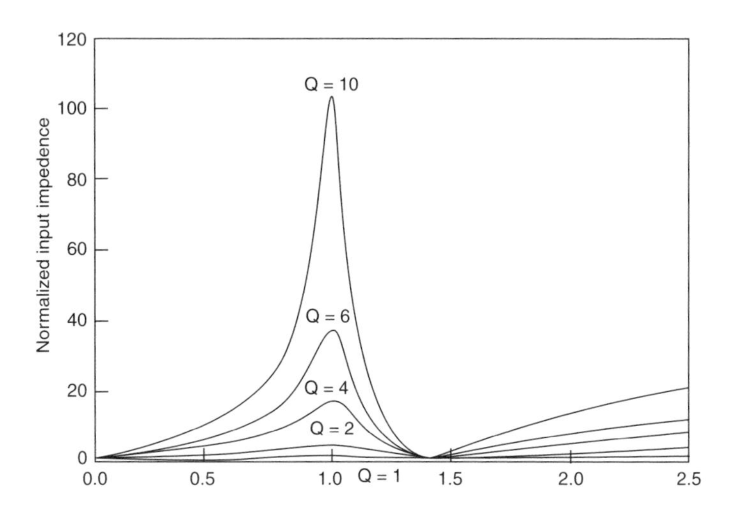

FIGURE 5.6
Input impedance $|Z_{in}(\beta)|$ vs. β referring to Q.

$$3\beta^4 - 8\beta^2 + 4 = 0 \tag{5.25}$$

It gives other two positive real roots as

$$\beta_3 = 0.816 \quad \beta_4 = 1.414$$

The former represents the switching frequency at which the local minimum gain is achieved, while at the latter the maximum voltage gain can be obtained.

From Equation (5.24), it should be noted that when $\beta = 1.414$ ($\beta^2 = 2$), the voltage gain $|g(\beta)|$ constantly keeps unity with any value of Q, as shown in Figure 5.4.

The absolute value of the input impedance $|Z_{in}(\beta)|$ and phase angle ϕ vs. β referring to various Q is shown in Figure 5.6 and Figure 5.7, respectively. The input impedance has its maximum value when the switching frequency is equal to natural resonant frequency (i.e., $\beta = 1$), and its minimum value when the $\beta = 1.414$. The phase angle ϕ keeps zero when $\beta = 1.414$, with any Q value, which means the resonant converter can be regarded as a pure resistive load.

5.2.3 Simulation and Experimental Results

Some simulation and experimental results shown in this section are helpful to understand the design and analysis. The chosen technical data are good examples to implement a particular RPC for reader's reference.

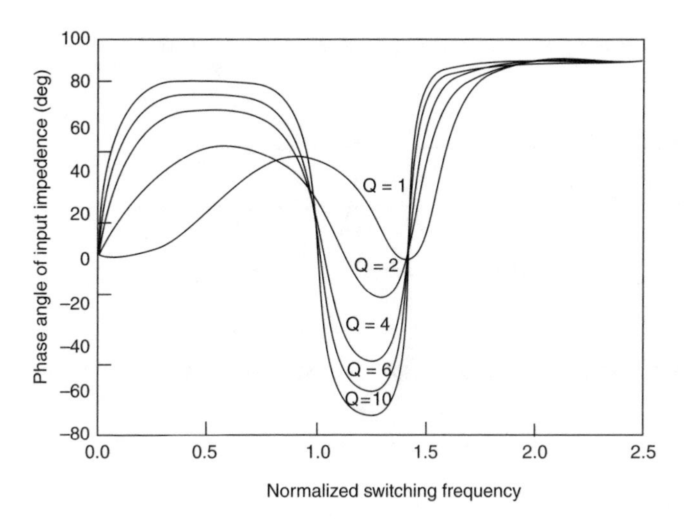

FIGURE 5.7
Phase angle ϕ vs. β referring to Q.

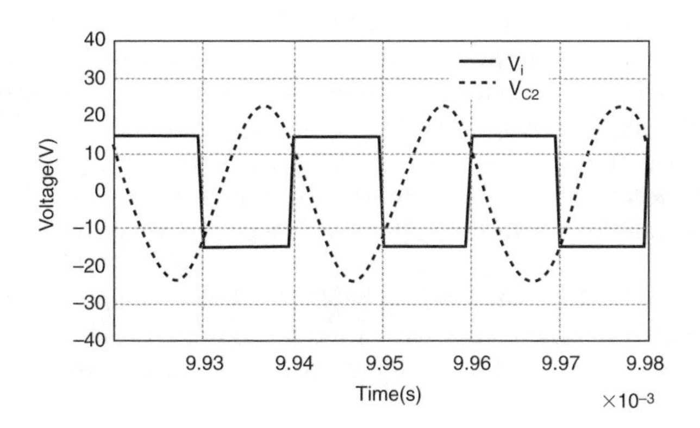

FIGURE 5.8
Simulation results at frequency f = 50 KHz.

5.2.3.1 Simulation Studies

In order to verify the mathematical analyses, a four-element cascade reverse double Γ-LC RPC is simulated using PSpice. The parameters used are $V_i = \pm 15\text{V}$, $L_1 = L_2 = 100\ \mu\text{H}$, $C_1 = C_2 = 0.22\ \mu\text{F}$, and $R = 22\ \Omega$. The natural resonant frequency is $f_0 = 1/\left(2\pi\sqrt{LC}\right) = 34$ kHz. The applying frequency is f = 50 KHz, corresponding to the $\beta = 1.42$. The simulation results are shown in Figure 5.8. The input signal V_i is a square waveform and the voltage across C_2 is a very smooth sinusoidal waveform.

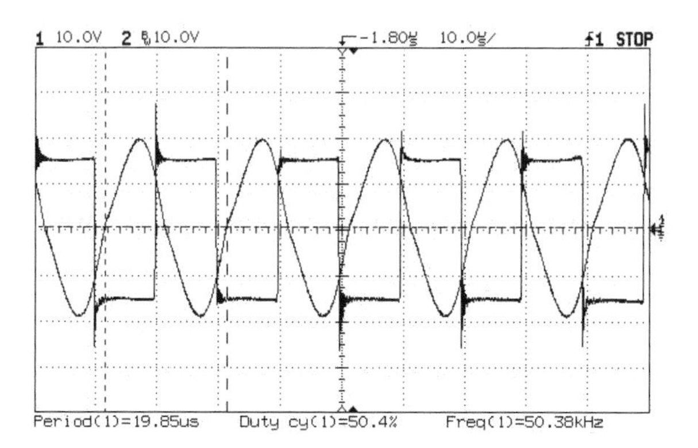

FIGURE 5.9
Experimental waveforms (f = 50.38 KHz).

5.2.3.2 Experimental Results

To verify the simulation results of the proposed cascade double Γ-LC RPC, a test rig was constructed with the same conditions: $V_i = \pm 15\,\text{V}$, $L_1 = L_2 =$ 100 μH, $C_1 = C_2 = 0.22$ μF and $R = 22\ \Omega$. The natural resonant frequency is $f_0 = 34$ kHz. For high frequency operation, the main power switch selected is IRF640 with its inner parasitic diode used as the antiparalleled diode. A high-speed integrated chip IR2104 is utilized to drive the half-bridge circuit. The switching frequency is chosen to be 50.38 KHz, corresponding to the $\beta = 1.42$. The experimental results are shown in Figure 5.9. The input signal V_i is a square waveform and the voltage across C_2 is a very smooth sinusoidal waveform.

Select the switching frequency to be 42.19 KHz, corresponding to the $\beta =$ 1.24. The experimental results are shown in Figure 5.10. The input signal V_i is a square waveform and the voltage across C_2 is also a smooth sinusoidal waveform.

5.3 Resonance Operation and Modeling

From the circuit diagram in Figure 5.1, the steady-state operation of the circuit is characterized by four operating modes within one switching period, when the resonant converter operates under continuous conduction mode (CCM). The equivalent circuits corresponding to each operating mode are depicted in Figure 5.11. Note that for a large output filter inductor, the bridge rectifier and the load can be represented as an alternating current sink with

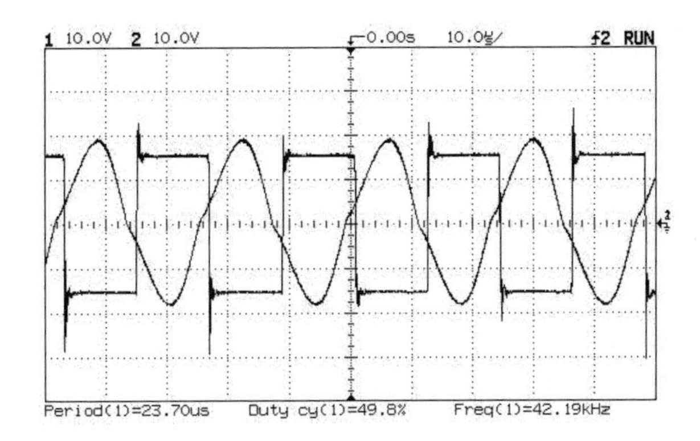

FIGURE 5.10
Experimental waveforms (f = 42.19 KHz).

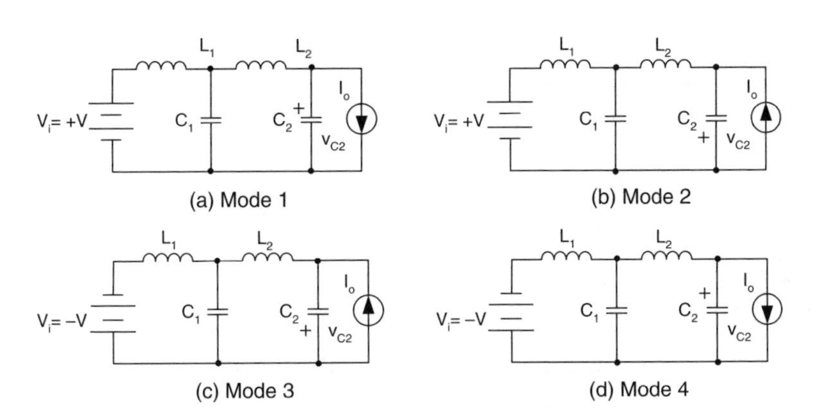

FIGURE 5.11
Different operating resonance modes.

constant amplitude I_0, synchronous with the polarity of the second resonant capacitor voltage v_{C2}.

5.3.1 Operating Principle, Operating Modes, and Equivalent Circuits

The operating modes of cascade reverse double Γ-LC RPC are very difficult to distinguish by analytic calculations, even under CCM conditions. However, the state of the converter should not exceed four modes when it operates above resonant frequency. The sequence among different modes is dependent on the phase angle to the voltage gain $V_{C2}(s)/V_i(s)$. In other words, when this angle is between 0° and +180°, the voltage across the second

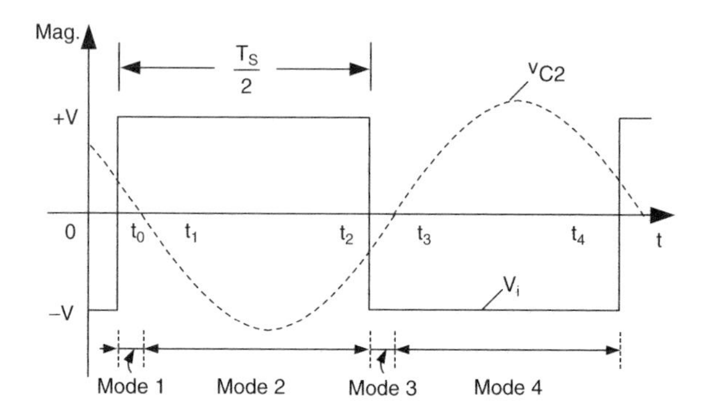

FIGURE 5.12
Voltage waveforms when v_{C2} leads V_i.

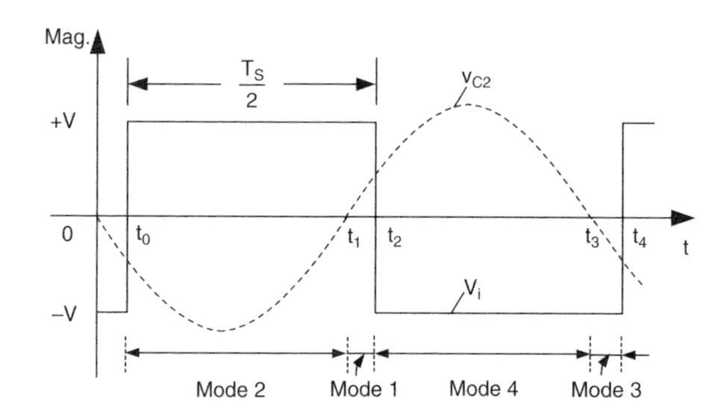

FIGURE 5.13
Voltage waveforms when v_{C2} lags behind V_i.

capacitor v_{C2} leads the input quasi-wave voltage V_i in Figure 5.12, thus the sequence from mode 1 to mode 4 should be:

- Mode 1 ($t_0 < t < t_1$): $V_i > 0$, $v_{C2} > 0$ (Figure 5.11a)
- Mode 2 ($t_1 < t < t_2$): $V_i > 0$, $v_{C2} < 0$ (Figure 5.11b)
- Mode 3 ($t_2 < t < t_3$): $V_i < 0$, $v_{C2} < 0$ (Figure 5.11c)
- Mode 4 ($t_3 < t < t_4$): $V_i < 0$, $v_{C2} > 0$ (Figure 5.11d)

Similarly, when the angle is between –180° and 0°, v_{C2} will lag behind the input voltage V_i in Figure 5.13, causing the sequence to be changed to:

- Mode 2 ($t_0 < t < t_1$): $V_i > 0$, $v_{C2} < 0$ (Figure 5.11b)
- Mode 1 ($t_1 < t < t_2$): $V_i > 0$, $v_{C2} > 0$ (Figure 5.11a)
- Mode 4 ($t_2 < t < t_3$): $V_i < 0$, $v_{C2} > 0$ (Figure 5.11d)
- Mode 3 ($t_3 < t < t_4$): $V_i < 0$, $v_{C2} < 0$ (Figure 5.11c)

In practice, when simulation is in progress, it is up to the algorithm that determines the shifting instant between different operating modes, by means of judging the switching period and the polarity of the second capacitor voltage.

5.3.2 State-Space Analysis

On the basis of the fact that the steady-state operation of resonant converter is periodic and composed of multiple operating modes, each mode stands for one state dependent on the different input voltage and rectifier current, thus, the state equation for each mode is given by:

$$\dot{x}_i = A_i x_i + B_i \tag{5.26}$$

where x_i is the state vector of the converter, A_i is the state coefficient matrix, B_i is the input vector of the converter in the ith operating mode, respectively. For i^{th} mode, Equation (5.26) can be solved analytically:

$$x_i(t) = e^{A_i t} x_i(t_0) + \int_0^t e^{A_i(t-\tau)} B_i d\tau = \Phi_i x_i(t_0) + \Gamma_i \tag{5.27}$$

where $\Phi_i = \Phi(t, t_0) = e^{A_i t}$ is the state transition matrix, $\Gamma_i = \int_0^t e^{A_i(t-\tau)} B_i d\tau$ and $x_i(t_0)$ are the initial conditions for the ith mode. For the continuous operation, each state solved in i^{th} mode will be employed as the initial conditions for the next $(I + 1)^{th}$ mode.

In fact, the solving process of Equation (5.27) is very tedious and time consuming by the requirement for evaluating the integral. However, by combining A_i and B_i to form an augmented dynamic matrix, the integration overhead can be eliminated at the expense of obtaining only the cyclic steady-state description:

$$\frac{d}{dt}\begin{pmatrix} x_i(t \\ 1 \end{pmatrix} = \begin{pmatrix} A_i & B_i \\ 0 & 0 \end{pmatrix}\begin{pmatrix} x_i(t) \\ 1 \end{pmatrix} \tag{5.28}$$

or

$$\frac{d}{dt}\hat{x}_i(t) = \hat{A}_i \hat{x}_i(t) \tag{5.29}$$

By means of the concept of state transition matrix, the solution for the state vector in different operating modes can be expressed as:

$$\hat{x}(t_1) = \hat{\Phi}_1 \hat{x}_1(t_0)$$

$$\hat{x}(t_2) = \hat{\Phi}_2 \hat{\Phi}_1 \hat{x}_1(t_0)$$

$$\ldots\ldots$$

For i^{th} mode:

$$\hat{x}(t_i) = \hat{\Phi}_i \hat{\Phi}_{i-1} \ldots \hat{\Phi}_1 \hat{x}(t_0) = \hat{\Phi}_{tot} \hat{x}(t_0) \tag{5.30}$$

where $\hat{\Phi}_i = \begin{pmatrix} \Phi_i & \Gamma_i \\ 0 & 1 \end{pmatrix}$ and $\hat{x}(t_i)$ is the state vector at time t_i.

Due to periodic nature of the system, the state vector at initial time t_0 should be equal to the one at final time t_i in one cycle, that is,

$$x(t_i) = \Phi_{tot} x(t_0) + \Gamma_{tot} = x(t_0) \tag{5.31}$$

yields:

$$x_{init}(t_0) = (I^n - \Phi_{tot})^{-1} \Gamma_{tot} \tag{5.32}$$

Thus, the state variables at any subsequent time are solved from Equation (5.30).

To obtain the average steady-state output voltage, the average values of the state-variables over a complete cycle are found from:

$$x_{av} = \frac{1}{T} \int_{t_0}^{t_0+T} x(t)dt \tag{5.33}$$

Again, the expression includes the integral process, which can be simplified by augmenting the state vector with

$$\dot{x}_{av}(t) = \frac{1}{T} x(t) \tag{5.34}$$

Then, consider the total dynamics of the converter during the ith mode, the simultaneous equations are given by:

$$\dot{x}_i = A_i x_i + B_i \tag{5.35}$$

$$\dot{x}_{i\_av} = \frac{d_i}{T} x_i \tag{5.36}$$

where d_i is the duty cycle for i^{th} mode in one cycle.

Substituting Equation (5.36) into Equation (5.28), the resulting dynamic equation is

$$\frac{d}{dt}\begin{pmatrix} x_i(t) \\ 1 \\ x_{i-av}(t) \end{pmatrix} = \begin{pmatrix} A_i & B_i & 0 \\ 0 & 0 & 0 \\ \frac{d_i}{T} I^n & 0 & 0 \end{pmatrix} \begin{pmatrix} x_i(t) \\ 1 \\ x_{i-av}(t) \end{pmatrix} \tag{5.37}$$

or

$$\dot{z}_i(t) = \tilde{A}_i z_i(t) \tag{5.38}$$

Again, as discussed in previous section, the state equation can be expressed as the function of state transition matrices in different modes:

$$z(t_i) = \tilde{\Phi}_i \tilde{\Phi}_{i-1} \cdots \tilde{\Phi}_1 z(t_0) \tag{5.39}$$

with the initial conditions:

$$z(t_0) = \begin{pmatrix} x_{init}(t_0) \\ 1 \\ 0 \end{pmatrix} \tag{5.40}$$

From the state variables, i.e., the voltage across the capacitors and current through the inductors, their average values can be gained directly.

The state coefficient matrix and the input source vector are determined in terms of different operating modes of the resonant converter. Considering the operating conditions described in Figure 5.12, for mode 1, suppose $V_I > 0$ and $v_{c2} > 0$, it can be obtained

$$A_1 = \begin{bmatrix} 0 & 0 & -1/L_1 & 0 & 0 & 0 \\ 0 & 0 & 1/L_2 & -1/L_2 & 0 & 0 \\ 1/C_1 & -1/C_2 & 0 & 0 & 0 & 0 \\ 0 & 1/C_2 & 0 & 0 & -1/C_2 & 0 \\ 0 & 0 & 0 & 1/L_f & 0 & -1/L_f \\ 0 & 0 & 0 & 0 & 1/C_f & -1/RC_f \end{bmatrix}$$

and
$$B_1 = [V/L_1 \quad 0 \quad 0 \quad 0 \quad 0 \quad 0]^T$$

where
$$x = (i_{L1}, i_{L2}, v_{C1}, v_{C2}, i_{Lf}, v_{Cf})^T$$

For mode 2, $V_i > 0$, $v_{C2} < 0$, the structure of the topology is changed due to the alteration of the polarity of the second capacitor voltage, giving:

$$A_2 = \begin{bmatrix} 0 & 0 & -1/L_1 & 0 & 0 & 0 \\ 0 & 0 & 1/L_2 & -1/L_2 & 0 & 0 \\ 1/C_1 & -1/C_2 & 0 & 0 & 0 & 0 \\ 0 & 1/C_2 & 0 & 0 & 1/C_2 & 0 \\ 0 & 0 & 0 & -1/L_f & 0 & -1/L_f \\ 0 & 0 & 0 & 0 & 1/C_f & -1/(RC_f) \end{bmatrix}$$

and
$$B_2 = B_1$$

For mode 3, $V_i < 0$, $v_{C2} < 0$, because only the input source is changed, the topology of the system remains invariant, thus the state coefficient matrix A_3 is identical with A_2, that is,

$$A_3 = A_2 \quad \text{and} \quad B_3 = [-V/L_1 \quad 0 \quad 0 \quad 0 \quad 0 \quad 0]^T$$

Similarly, when operating the above resonance, for mode 4, $V_i < 0$, $v_{C2} > 0$, yielding:

$$A_4 = A_1 \quad \text{and} \quad B_4 = B_3$$

By employing the above coefficient matrices and vectors with the initial conditions, the values of all state variables in different operating modes can be obtained following Equation (5.39). Thus, the dynamic operating behavior of the resonant converter can be described by these state variables as well.

5.4 Small-Signal Modeling of Cascade Reverse Double Γ-LC RPC

In the previous section, state-space averaging technique has been applied to investigate the dynamic behavior and successfully simulated the waveforms

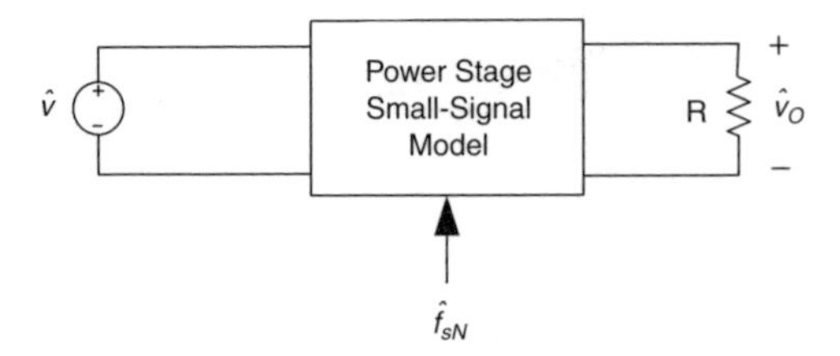

FIGURE 5.14
Small-signal perturbations in input voltage and switching frequency.

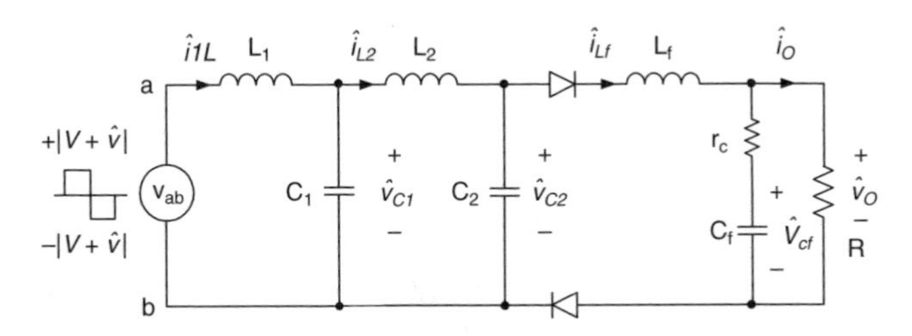

FIGURE 5.15
Equivalent circuit of cascade reverse double Γ-LC RPC.

at different time intervals, however, the numerical results cannot reveal the relations of various control specialties, e.g., frequency response, closed-loop control system stability, etc. In order to study these characteristics more deeply, a number of mathematical methods were presented among them the small-signal modeling is implemented.

5.4.1 Small-Signal Modeling Analysis

5.4.1.1 *Model Diagram*

The block diagram of the small-signal model is depicted in Figure 5.14, where $\hat{v}$ and $\hat{f}_{sN}$ represent small-signal perturbation of the line voltage and the frequency control signal. The output variable is the perturbed output voltage, $\hat{v}_o$. With the model, it is easy to obtain the commonly used small-signal transfer functions, such as control-to-output transfer function, line-to-output transfer function, input impedance, and output impedance.

5.4.1.2 *Nonlinear State Equation*

The equivalent circuit of the resonant converter is shown in Figure 5.15. As seen, the half-bridge circuit employing two diodes, applies a square-wave voltage, v_{ab}, to the resonant network. Suppose the converter operates above resonance, the state equations of the resonant converter can be obtained, where the nonlinear terms are in bold face:

$$\begin{aligned}
&L_1 \frac{di_{L1}}{dt} + v_{C1} = \boldsymbol{v}_{ab} \\
&C_1 \frac{dv_{C1}}{dt} + i_{L2} = i_{L1} \\
&L_2 \frac{di_{L2}}{dt} + v_{C2} = v_{C1} \\
&C_2 \frac{dv_{C2}}{dt} + \boldsymbol{sgn(v_{C2})i_{Lf}} = i_{L2} \\
&L_f \frac{di_{Lf}}{dt} + (i_{Lf} - i_o)r_c + v_{Cf} = |\boldsymbol{v}_{C2}| \\
&C_f \frac{dv_{Cf}}{dt} = i_{Lf} - i_o
\end{aligned} \tag{5.41}$$

The output variable is the output voltage, v_o, which gives:

$$v_o = (i_{Lf} - i_o)r_c + v_{Cf} \tag{5.42}$$

In this circuit, the output voltage is regulated either by the input line voltage, v, or by the applying switching frequency, ω. Thus, the operating point $\mathcal{P}$ can be expressed as the function of these variables $\mathcal{P} = \{v, R, \omega\}$.

5.4.1.3 Harmonic Approximation

Under the assumption that both the voltage and current inside the resonant network are quasi-sinusoidal, the so-called fundamental approximation method is applied to the derivation of the small-signal models. In other words, the variables in the resonant network are assumed as:

$$\begin{aligned}
&i_{L1} = i_{1s}(t)\sin\omega t + i_{1c}(t)\cos\omega t \\
&i_{L2} = i_{2s}(t)\sin\omega t + i_{2c}(t)\cos\omega t \\
&v_{C1} = v_{1s}(t)\sin\omega t + v_{1c}(t)\cos\omega t \\
&v_{C2} = v_{2s}(t)\sin\omega t + v_{2c}(t)\cos\omega t
\end{aligned} \tag{5.43}$$

Note that the envelope terms $\{i_{1s}, i_{1c}, i_{2s}, i_{2c}, v_{1s}, v_{1c}, v_{2s}, \mathrm{v}_{2c}\}$ are slowly time varying, thus the dynamic behavior of these terms can be investigated. The derivatives of i_{L1}, i_{L2}, v_{C1} and v_{C2} are found to be:

$$\frac{di_{L1}}{dt}=[\frac{di_{1s}}{dt}-\omega i_{1c}]\sin\omega t+[\frac{di_{1c}}{dt}+\omega i_{1s}]\cos\omega t$$
$$\frac{di_{L2}}{dt}=[\frac{di_{2s}}{dt}-\omega i_{2c}]\sin\omega t+[\frac{di_{2c}}{dt}+\omega i_{2s}]\cos\omega t$$
$$\frac{dv_{C1}}{dt}=[\frac{dv_{1s}}{dt}-\omega v_{1c}]\sin\omega t+[\frac{dv_{1c}}{dt}+\omega v_{1s}]\cos\omega t \tag{5.44}$$
$$\frac{dv_{C2}}{dt}=[\frac{dv_{2s}}{dt}-\omega v_{2c}]\sin\omega t+[\frac{dv_{2c}}{dt}+\omega v_{2s}]\cos\omega t$$

5.4.1.4 Extended Describing Function

By employing the extended describing function modeling technique stated in the literature, the nonlinear terms in Equation (5.41) can be approximated either by the fundamental harmonic terms or by the DC terms, to yield:

$$v_{ab}(t)\approx f_1(v)\sin\omega_s t$$
$$sgn(v_2)i_{Lf}\approx f_2(v_{2s},v_{2c},i_{Lf})\sin\omega_s t+f_3(v_{2s},v_{2c},i_{Lf})\cos\omega_s t \tag{5.45}$$
$$|v_{C2}|\approx f_4(v_{2s},v_{2c})$$

These functions are called extended describing functions (EDFs). They are dependent on the operating conditions and the harmonic coefficients of the state variables. The EDF terms can be calculated by making Fourier expansions of the nonlinear terms, to give:

$$f_1(v)=\frac{4}{\pi}v$$
$$f_2(v_{2s},v_{2c},i_{Lf})=\frac{4}{\pi}\frac{v_{2s}}{A_p}i_{Lf}$$
$$f_3(v_{2s},v_{2c},i_{Lf})=\frac{4}{\pi}\frac{v_{2c}}{A_p}i_{Lf} \tag{5.46}$$
$$f_4(v_{2s},v_{2c})=\frac{2}{\pi}A_p$$

where

$$A_p=\sqrt{v_{2s}^2+v_{2c}^2}$$

is the peak voltage of the second capacitor voltage v_{C2}.

5.4.1.5 Harmonic Balance

Substituting Equation (5.43) to Equation (5.46) into Equation (5.41), the nonlinear large-signal model of cascade reverse double Γ-LC RPC is obtained as follows:

$$
\begin{aligned}
&L_1(\frac{di_{1s}}{dt} - \omega_s i_{1c}) + v_{1s} = \frac{4}{\pi} v \\
&L_1(\frac{di_{1c}}{dt} + \omega_s i_{1s}) + v_{1c} = 0 \\
&C_1(\frac{dv_{1s}}{dt} - \omega_s v_{1c}) = i_{1s} - i_{2s} \\
&C_1(\frac{dv_{1c}}{dt} + \omega_s v_{1s}) = i_{1c} - i_{2c} \\
&L_2(\frac{di_{2s}}{dt} - \omega_s i_{2c}) = v_{1s} - v_{2s} \\
&L_2(\frac{di_{2c}}{dt} + \omega_s i_{2s}) = v_{1c} - v_{2c} \\
&C_2(\frac{dv_{2s}}{dt} - \omega_s v_{2c}) + \frac{4}{\pi}\frac{v_{2s}}{A_p} i_{Lf} = i_{2s} \\
&C_2(\frac{dv_{2c}}{dt} + \omega_s v_{2s}) + \frac{4}{\pi}\frac{v_{2c}}{A_p} i_{Lf} = i_{2c} \\
&L_f \frac{di_{Lf}}{dt} + (i_{Lf} - i_o)r_c = \frac{2}{\pi} A_p - v_{Cf} \\
&C_f \frac{dv_{Cf}}{dt} = i_{Lf} - i_o
\end{aligned}
\tag{5.47}
$$

The corresponding output equation is

$$v_o = (i_{Lf} - i_o)r_c + v_{Cf} \tag{5.48}$$

It should be noted that the small-signal modulation frequency is lower than the switching frequency, thus the nonlinear model can be linearized by perturbing the system around the operating point $\mathcal{P}$. The perturbed variables are the inputs, the state variables, and the outputs. Each one has the form of

$$x(t) = X + \hat{x}(t)$$

where X is the steady state at the operating point, and $\hat{x}(t)$ is a small amplitude perturbation.

Similarly, in this circuit, the input variables are found to be

$$v = V + \hat{v} \qquad \omega_s = \Omega_s + \hat{\omega}_s$$

5.4.1.6 Perturbation and Linearization

Under the small amplitude perturbation assumptions, the complete linearized small-signal models can be established by applying the perturbation to Equation (5.47) and only considering the first partial derivatives, to give:

$$\begin{aligned}
&L_1 \frac{d\hat{i}_{1s}}{dt} = Z_{L1}\hat{i}_{1c} + E_{1s}\hat{f}_{sN} - \hat{v}_{1s} + k_v\hat{v} \\
&L_1 \frac{d\hat{i}_{1c}}{dt} = -Z_{L1}\hat{i}_{1s} - E_{1c}\hat{f}_{sN} - \hat{v}_{1c} \\
&C_1 \frac{d\hat{v}_{1s}}{dt} = \hat{i}_{1s} - \hat{i}_{2s} + G_s\hat{v}_{1c} + J_{1s}\hat{f}_{sN} \\
&C_1 \frac{d\hat{v}_{1c}}{dt} = \hat{i}_{1c} - \hat{i}_{2c} - G_s\hat{v}_{1s} - J_{1c}\hat{f}_{sN} \\
&L_2 \frac{d\hat{i}_{2s}}{dt} = \hat{v}_{1s} - \hat{v}_{2s} + Z_{L2}\hat{i}_{2c} + E_{2s}\hat{f}_{sN} \\
&L_2 \frac{d\hat{i}_{2c}}{dt} = \hat{v}_{1c} - \hat{v}_{2c} - Z_{L2}\hat{i}_{2s} - E_{2c}\hat{f}_{sN} \\
&C_2 \frac{d\hat{v}_{2s}}{dt} = \hat{i}_{2s} - g_{ss}\hat{v}_{2s} + g_{sc}\hat{v}_{2c} - 2k_s\hat{i}_{Lf} + J_{2s}\hat{f}_{sN} \\
&C_2 \frac{d\hat{v}_{2c}}{dt} = \hat{i}_{2c} - g_{cc}\hat{v}_{2c} + g_{cs}\hat{v}_{2s} - 2k_c\hat{i}_{Lf} + J_{2c}\hat{f}_{sN} \\
&L_f \frac{d\hat{i}_{Lf}}{dt} = (\hat{i}_o - \hat{i}_{Lf})r_c - \hat{v}_{Cf} + k_s\hat{v}_{2s} + k_c\hat{v}_{2c} \\
&C_f \frac{d\hat{v}_{Cf}}{dt} = \hat{i}_{Lf} - \hat{i}_o
\end{aligned} \tag{5.49}$$

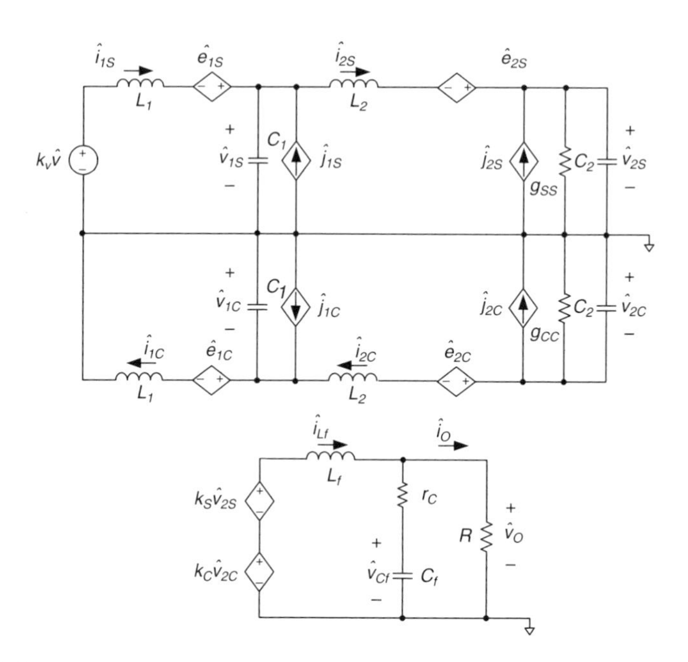

FIGURE 5.16
Equivalent small-signal circuit of cascade reverse double Γ-LC RPC.

where the input variables are $\hat{v}$ and $\hat{f}_{sN}$, standing for the perturbed line voltage and normalized switching frequency, respectively.

The output part of the small-signal model is given by:

$$\hat{v}_o = (\hat{i}_{Lf} - \hat{i}_o)r_c + \hat{v}_{Cf} \tag{5.50}$$

All the parameters used in the above models are given in the Appendix.

5.4.1.7 Equivalent Circuit Model

This linearized small-signal model makes it possible to describe the operating characteristics of the resonant converter using equivalent circuit model, as shown in Figure 5.5. In this model, the circuit is divided into two parts: the resonant network and the output part. To simplify the drawing, some dependent sources are defined as:

$$\hat{e}_{1s} = Z_{L1}\hat{i}_{1c} + E_{1s}\hat{f}_{sN} \qquad \hat{e}_{1c} = Z_{L1}\hat{i}_{1s} + E_{1c}\hat{f}_{sN}$$

$$\hat{e}_{2s} = Z_{L2}\hat{i}_{2c} + E_{2s}\hat{f}_{sN} \qquad \hat{e}_{2c} = Z_{L2}\hat{i}_{2s} + E_{2c}\hat{f}_{sN}$$

$$\hat{j}_{1s} = G_s\hat{v}_{1c} + J_{1s}\hat{f}_{sN} \qquad \hat{j}_{1c} = G_s\hat{v}_{1s} + J_{1c}\hat{f}_{sN}$$

$$\hat{j}_{2s} = g_{sc}\hat{v}_{2c} - 2k_s\hat{i}_{Lf} + J_{2s}\hat{f}_{sN} \qquad \hat{j}_{2c} = g_{cs}\hat{v}_{2s} - 2k_c\hat{i}_{Lf} + J_{2c}\hat{f}_{sN}$$

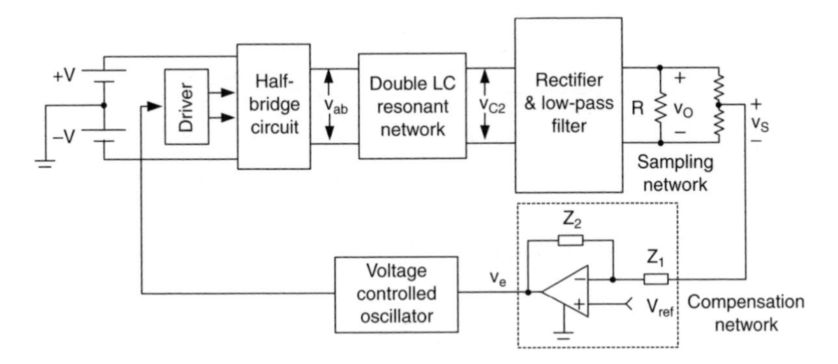

FIGURE 5.17
Closed-loop system of cascade reverse double Γ-LC RPC.

By employing the Kirchoff's current law and voltage law, the equivalent circuit can be drawn following the state-equation Equation (5.49). Note that in the output part, the voltage across the second capacitor $|\hat{v}_{C2}|$ is replaced by two voltage-controlled sources $\{k_s\hat{v}_{2s}, k_c\hat{v}_{2c}\}$. This circuit model can be implemented in general-purpose simulation software, such as PSpice or MATLAB, to obtain the frequency response of the system.

5.4.2 Closed-Loop System Design

The closed-loop control system diagram of the half-bridge cascade reverse double Γ-LC RPC is illustrated in Figure 5.17, where the feedback loop is composed of the sampling network, the compensator, and the voltage-controlled oscillator (VCO). The sampling network, R_1 and R_2, contributes attenuation according to its sampling ratio of $R_1/(R_1 + R_2)$. Since the two resistors are chosen to be equal, the gain attenuation of the network is 20 log (v_s/v_o) = –6dB. The sampled voltage, v_s, is then sent to the inverting input side of the error amplifier, where it is compared with a fixed reference voltage v_{ref} and generates an error voltage, v_e. This voltage determines the frequency output of a voltage-controlled oscillator (VCO), whose gain can be obtained either in datasheet or by experiment. The Bode plot, considering all the voltage gains of the main circuit, the sampling network and VCO, is shown in Figure 5.18.

It should be noted that the slope of the magnitude response at the unity-gain crossover frequency is –40dB/decade. Both the gain margin and the phase margin of the small-signal model are not able to meet the requirements of the stability. Thus, the alleged feedback compensation is often used to shape the frequency response such that it remains stable under all operating conditions, especially in the presence of noise or disturbance injected at any point in the loop. In order to yield a –20dB/decade slope at the unity-gain crossover frequency, the magnitude response of the compensation network must have a slope of +20dB/decade slope at the unity-gain crossover frequency. Hence, a three poles and double zeros compensation network is

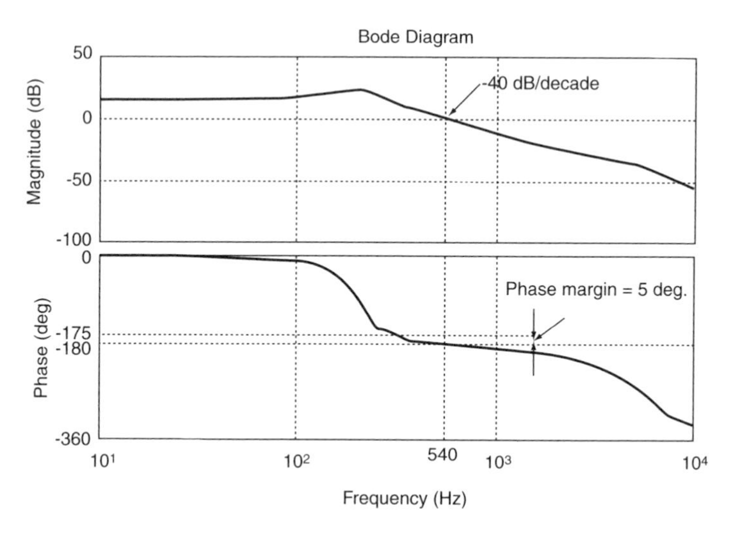

FIGURE 5.18
Bode diagram of the small-signal equivalent circuit.

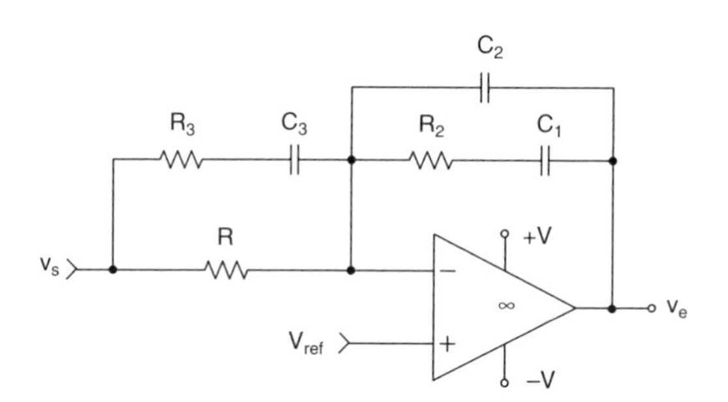

FIGURE 5.19
Compensation network.

employed, whose circuit implementation and Bode diagram are shown in Figure 5.19 and Figure 5.20 respectively.

The transfer function for this compensation network is

$$H(j\omega) = \frac{1 + j\omega R_2 C_1}{-\omega^2 R_2 C_1 C_2 + j\omega(C_1 + C_2)} \frac{1 + j\omega(R_1 + R_3)C_3}{R_1 + j\omega R_1 R_3 C_3} \tag{5.51}$$

As seen from Equation (5.51), it has two high-frequency poles, one at $f_{p1} = 1/2R_3C_3$ and the other at $f_{p2} = (C_1 + C_2)/2R_2C_1C_2$. The zeros are at $f_{z1} = 1/2R_2C_1$ and $f_{z2} = 1/2(R_1 + R_3)C_3$, respectively. The two gains of the compensation network are $K_1 = R_2/R_1$ and $K_2 = R_2(R_1 + R_3)/R_1R_3$, respectively. To simplify the design process, the two high frequency poles are usually chosen to be equal to each other (i.e., $f_{p1} = f_{p2} = f_p$) such that

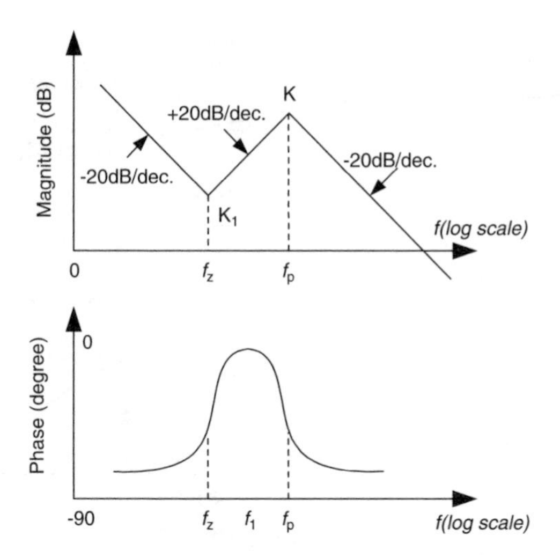

FIGURE 5.20
Bode schematic diagram of the compensation network.

$$\frac{1}{2\pi R_3 C_3} = \frac{C_1 + C_2}{2\pi R_2 C_1 C_2} \tag{5.52}$$

The phase lag due to the double poles, θ_p, is

$$\theta_p = 2\tan^{-1}(\frac{f_p}{f_1}) \tag{5.53}$$

where f_1 is the unity-gain crossover frequency.
Similarly, the two zeros are also chosen to be equal, such that

$$\frac{1}{2\pi R_2 C_1} = \frac{1}{2\pi (R_1 + R_3) C_3} \tag{5.54}$$

The phase boost at the double zeros, θ_z, is

$$\theta_z = 2\tan^{-1}(\frac{f_1}{f_z}) \tag{5.55}$$

Hence the total phase lag introduced by the compensation network and the error amplifier at the unity-gain crossover frequency is

$$\theta_c = 270° - 2\tan^{-1}(\frac{f_1}{f_z}) + 2\tan^{-1}(\frac{f_p}{f_1}) \tag{5.56}$$

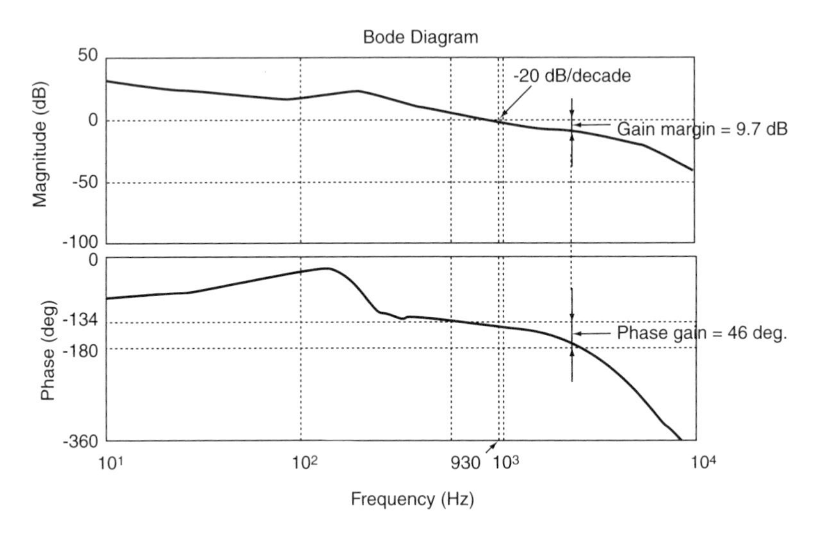

FIGURE 5.21
Bode diagram of the open loop system.

where the 270° phase lag is due to the phase inversion introduced by the inverting amplifier and the pole at the origin of the compensation network.

Consider the real system described in Figure 5.21, the unity-gain crossover frequency is chosen to be 900 Hz, where the attenuation is –16 dB. Hence, the gain of the error amplifier at the unity-gain crossover frequency is chosen to be +16 dB in order to yield 0 dB at the unity-gain crossover frequency.

The locations of the double poles and double zeros of the compensation network are chosen to yield the desired phase margin of 45°. The total phase shift at the unity-gain crossover frequency is 360° – 45° or 315°. Taking into account the effect of equivalent series resistor (ESR) in the output capacitor, the phase lag of the output filter with an output capacitor ESR is

$$\theta_{LC} = 180° - \tan^{-1}(\frac{f_1}{f_{ESR}}) \tag{5.57}$$

where f_{ESR} is the ESR break frequency

$$f_{ESR} = \frac{1}{2\pi r_c C_f} \tag{5.58}$$

Then the phase lag contribution from the compensation network and the error amplifier is

$$\theta_{ea} = 315° - \theta_{LC} \tag{5.59}$$

Hence, from Equation (5.58), the phase lag contribution from the compensation network is

$$2\tan^{-1}(\frac{3}{f_z}) - 2\tan^{-1}(\frac{f_p}{3}) = 270° - \theta_{ea} \tag{5.60}$$

Solving this equation yields a value of 4.64 to achieve a phase lag of 65.67°. Hence, the high-frequency pole should be located at 4.64 times the unity-gain crossover frequency, or 4.2 KHz, while the low-frequency zero should be located at one-fourth of the unity-gain crossover frequency, or 190 Hz. There are six components to be selected for the compensation network. As described previously, the gain at the double zero, K_1, is 0 dB, or 1. Assuming an R_1 value of 1 KΩ, R_2 is 1 KΩ too. The gain at the double poles, K_2, is selected to be 20 dB, or 10. Thus, *R*3 is given by

$$R_3 = \frac{R_1 R_2}{K_2 R_1 - R_2} \tag{5.61}$$

The capacitance value for C_3 is

$$C_3 = \frac{1}{2\pi f_p R_3} \tag{5.62}$$

Hence, the capacitance values of C_1 and C_2 can be calculated respectively from Equation (5.52) and Equation (5.54).

The Bode diagram of the open loop system is given in Figure 5.21. Notice that the gain at the unity-gain crossover frequency in the overall magnitude response of the feedback-compensated resonant converter is 0 dB with a slope of –20 dB/decade. The unity-gain crossover frequency is enhanced from 540 KHz to 930 KHz, at which point the phase lag is reduced from –175° to –134°. The values of gain margin and phase margin are 9.7 dB and 46°, respectively. Both of them have fit the specified requirements of the stability.

In most cases, the stability plays an important role in various performance indexes of a closed-loop system. For this resonant converter, the stability can be studied by analyzing the characteristics of the poles and zeros of the closed-loop system. Since the small-signal model has been found in previous section, the state-space equations can be established from Equation (5.49), to give:

$$\frac{d\hat{x}}{dt} = A\hat{x} + B\hat{u}$$

and

$$\hat{y} = C\hat{x} + D\hat{u}$$

where

$$\hat{x} = \left(\hat{i}_{1s}, \hat{i}_{1c}, \hat{v}_{1s}, \hat{v}_{1c}, \hat{i}_{2s}, \hat{i}_{2c}, \hat{v}_{2s}, \hat{v}_{2c}, \hat{i}_{Lf}, \hat{v}_{Cf}\right)^T$$

$$\hat{u} = \left(\hat{v}, \hat{f}_{sN}, \hat{i}_o\right)^T \qquad \hat{y} = \hat{v}_o$$

Based on the relationship between the state-space equation and transfer function, these equations of the open loop system can be transformed to the transfer function $G_1(s)$, to give:

$$G_1(s) = \frac{\hat{Y}(s)}{\hat{U}(s)} = C[sI - A]^{-1}B + D \tag{5.63}$$

By considering both the expression in Equation (5.51) and $G_1(s)$, the closed-loop control system transfer function can be obtained. Hence, the poles of the closed-loop control system are found to be:

$p_{1,2} = -0.3138 \pm 19.3296i \qquad p_{3,4} = -0.8750 \pm 12.9716i$

$p_{5,6} = -0.8809 \pm 5.2819i \qquad p_{7,8} = -0.3914 \pm 0.9492i$

$p_{9,10} = -0.0071 \pm 0.0377i \qquad p_{11,12} = -0.7777 \pm 0.0000i$

$p_{13} = 0$

Notice that all the real parts of the poles are nonnegative, thus, the whole system is found stable. The root loci of the closed-loop system are given in Figure 5.22.

5.5 Discussion

In all the previous sections, the analyses are undertaken on the basis of the assumption that two resonant inductors are identical, as well as two resonant capacitors. The following discussion will be focused on the condition that the resonant components have different values, that is, variable-parameters.

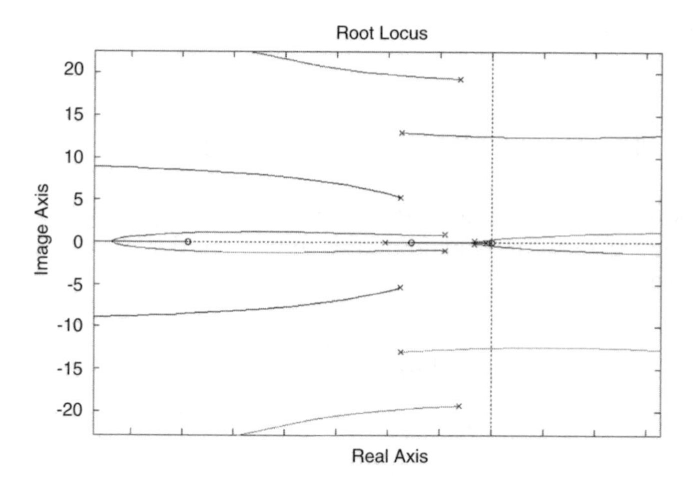

FIGURE 5.22
Root loci of the closed-loop system.

In addition, the discontinuous operation, always occurring when switching frequency is lower than the natural resonant frequency, will also be taken into consideration.

5.5.1 Characteristics of Variable-Parameter Resonant Converter

In fact, most of the derivation results and general conclusions in Section 5.3 and Section 5.4 are still valid for the variable-parameter condition, except that the curves of the voltage transfer gain are a little bit different. For instance, if the ratios of two inductors and two capacitors are defined as:

$$p = L_1 / L_2 \quad \text{and} \quad q = C_1 / C_2$$

then the expression in Equation (5.1) can be rewritten as :

$$g(\omega) = \frac{R_{eq}}{\begin{pmatrix} \omega^4 p\beta L^2 C^2 R_{eq} - \omega^2 (pq + p + 1) LCR_{eq} \\ + R_{eq} + j[\omega L(p+1) - \omega^3 L^2 Cpq] \end{pmatrix}} \tag{5.64}$$

Following the same definitions of the natural resonant frequency ω_0, Q value, and relative frequency β, the voltage transfer gain $g(\beta)$ is given by:

$$g(\beta) = \frac{1}{1 - \beta^2 (pq + p + 1) + \beta^4 pq + j[(p+1) - pq\beta^2]\beta Q} \tag{5.65}$$

and its determinant is

$$|g(\beta)| = \frac{1}{\sqrt{[1-\beta^2(pq+p+1)+\beta^4 pq]^2 + [(p+1)-pq\beta^2]^2\beta^2 Q^2}} \tag{5.66}$$

The local maximum and minimum values of the voltage gain $g(\beta)$ can be gained by solving the resulting equation after setting the derivatives of the determinant to zero.

$$\frac{d}{d\beta^2}|g(\beta)| = 0 \tag{5.67}$$

$$\begin{aligned} &4p^2q^2\beta^6 + [3p^2q^2Q^2 - 6pq(pq+p+1)]\beta^4 \\ &+ [4pq + 2(pq+p+1)^2 - 4pq(p+1)Q^2]\beta^2 \\ &+ Q^2(p+1)^2 - 2(pq+p+1) = 0 \end{aligned} \tag{5.68}$$

When $p = q = 1$, this equation is simplified to be

$$4\beta^6 + \left(3Q^2 - 18\right)\beta^4 + \left(22 - 8Q^2\right)\beta^2 + 4Q^2 - 6 = 0$$

which is the same as Equation (5.21).

Figure 5.23a to d depict the voltage gain $|g(\beta)|$ with different parameter ratios referring to various Q values. One can find that all the curves have the similar shape in Figure 5.4, which begins with unity at low switching frequency and displays two peaks with lower Q values. However, the different parameter ratio results in a different bandwidth. Thus, the designer will have more choice to find the most appropriate bandwidth to meet the requirements of various applications. Besides this, with certain parameter ratios (for example, $p = 2$, $q = 1$), all the curves intersect at one point, which is independent of the Q values. The corresponding switching frequency can be calculated by setting the imaginary part of the denominator in the expression in Equation (5.65) to zero, gives:

$$p + 1 = pq\tilde{\beta}^2$$

or

$$\tilde{\beta}^2 = \frac{p+1}{pq} \tag{5.69}$$

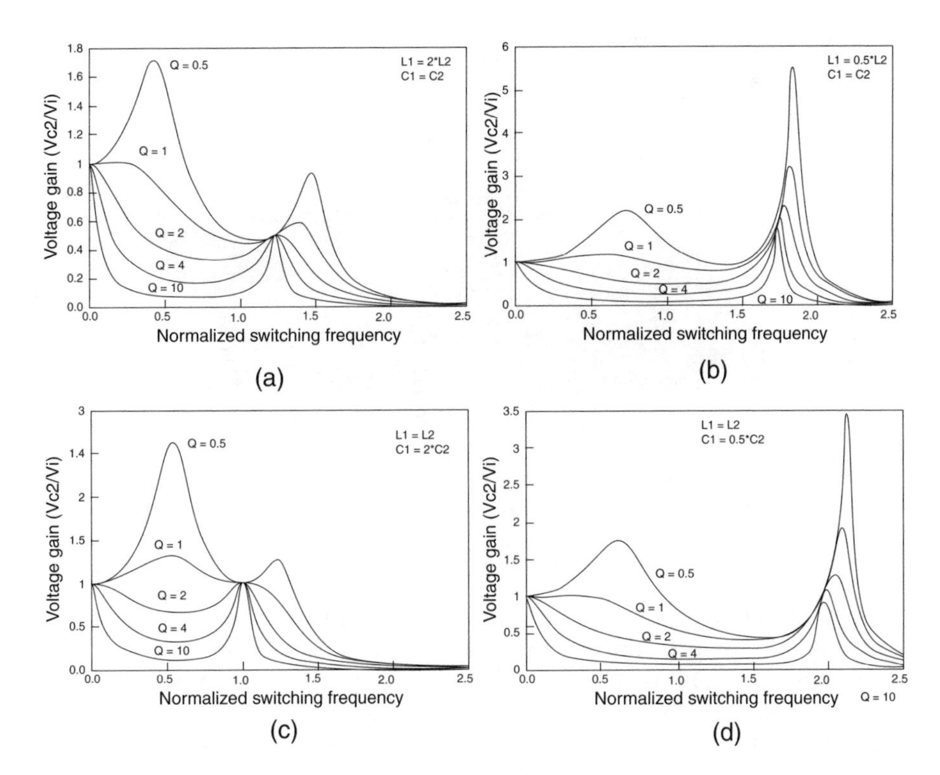

FIGURE 5.23
Voltage gain $|g(\beta)|$ with different parameter ratios referring to various Q.

Substituting Equation (5.69) into Equation (5.65), the voltage gain at this frequency is obtained as:

$$|g(\tilde{\beta})| = \frac{1}{p} \tag{5.70}$$

Hence, the voltage transfer gain is only dependent on the inductors ratio α and it determines the rough shape of the gain curve over a wide frequency range $[0 \sim \tilde{\beta}]$, which is useful for the designer to estimate the needed operating point.

Figure 5.24 gives the maximum voltage transfer gain vs. different parameter ratios of p and q, corresponding to the peak value in Figure 5.23a to d. Notice that the maximum gain is obtained with small p and large q. This is true because when L_1 is much smaller than L_2 while C_1 is larger than C_2, the cascade reverse double Γ-LC RPC will be degraded to the conventional parallel resonant converter, which voltage gain is prominent near the natural resonant frequency. Figure 5.24 shows the three-dimensional relations.

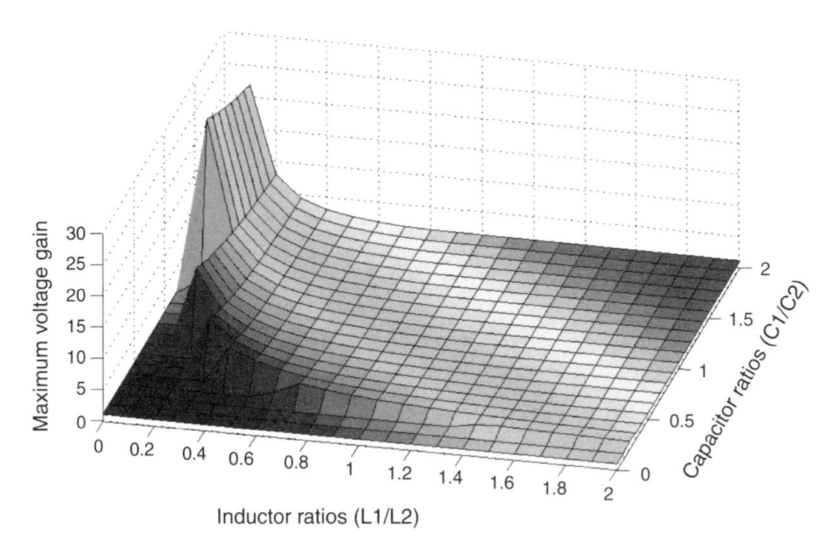

FIGURE 5.24
Maximum voltage gain vs. different parameter ratios of p and q.

The dynamic analysis of the variable-parameter resonant converter is similar to that in the previous section, except that some relevant matrices need to be amended. The waveforms of the input and the resonant voltage are shown in Figure 5.25a, where some distortions can be found in the second capacitor voltage v_{c2}. Figure 5.25b is the probe output of the commonly used circuit simulation software PSpice. From this one can find a good accordance between the two figures.

The comparison of the resonant output voltage and total harmonic distortion (THD) of the variable-parameter resonant converter is given in Table 5.1. Compared to other conditions, the second capacitor voltage obtains the highest amplitude when $L_1 = L_2$, $C_1 = C_2$, while it keeps the lowest total harmonic distortion value. Thus, it is reasonable to implement the practical cascade reverse double Γ-LC RPC with dual symmetric resonant network.

5.5.2 Discontinuous Conduction Mode (DCM)

When the switching frequency is lower than half of the natural resonant frequency, the current through the inductors and the voltage across the capacitors will become discontinuous. For the cascade reverse double Γ-LC PRC, in DCM operation, the energy in the resonant capacitor is consumed before the new half of a switching cycle. There are six stages of operation in one switching cycle, as shown in Figure 5.26. Suppose before the beginning of a switching cycle, the second inductor current i_{L2} is zero. When stage 1 begins at time t_0, since i_{L2} is lower than the constant-current-sink I_o, all the rectifier diodes become forward biased and conduct. The second capacitor

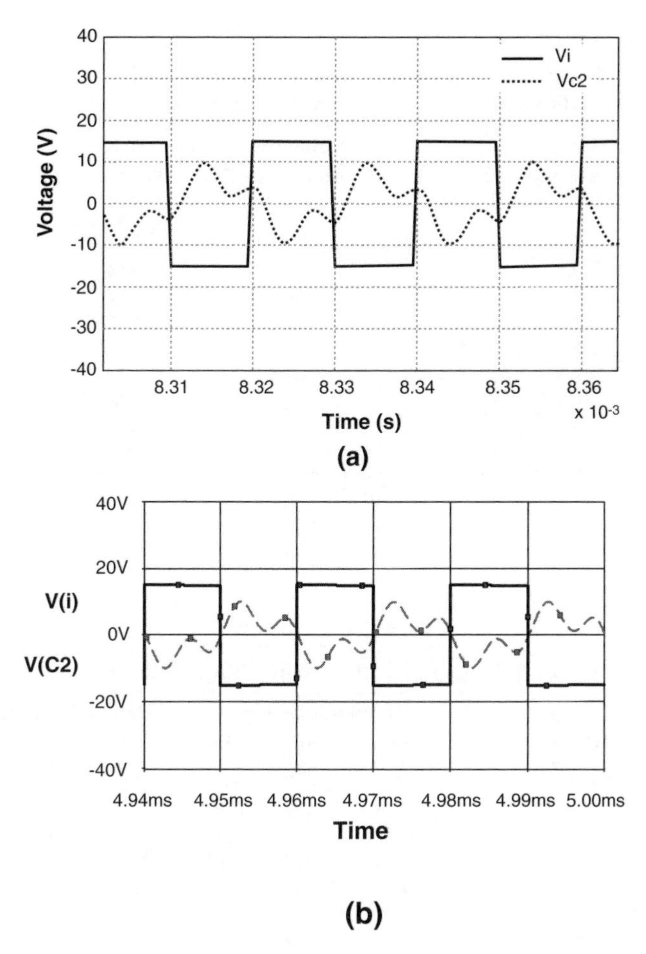

FIGURE 5.25
Simulation waveforms using Matlab (a) and PSpice (b) when $L_1 = 10\ L_2$.

C_2 is clamped to zero volts by the freewheeling action, which is shown in Figure 5.27a. Thus, the second inductor L_2 is charged and the current though it is increased. Stage 2 commences when the current i_{L2} reaches the magnitude of the current-sink I_0, at time t_1. The topology at this moment is the same as that in CCM operation. Stage 3 starts when the input voltage source V_i changes its polarity at time t_2, as depicted in Figure 5.27c. This completes the first half-cycle of operation. The next half-cycle operation repeats the same way as the first half-cycle, except that the direction of the inductor current i_{L2} and the polarity of the capacitor voltage v_{C2} are reversed.

The state coefficient matrix and input source vector corresponding to two additional operating modes are given by:

TABLE 5.1

Comparison of Output Resonant Voltage and THD Values under Different Parameter Ratios

		Switching frequency f_s = 50 KHz	
$\alpha(L_1/L_2)$	$\beta(C_1/C_2)$	V_{c2} Amplitude (V)	THD (%)
1	1	23.5	2.41
2	1	11.0	7.90
0.5	1	17.0	6.84
1	2	6.0	5.82
1	0.5	8.0	5.93
2	2	3.5	2.72
0.5	0.5	10.0	7.14
0.5	2	19.0	8.55
2	0.5	6.0	3.91

Note: With L_2 = 100 μH, C_2 = 0.22 μF, R = 22 Ω, L_f = 2.4 mH, C_f = 220 μF

TABLE 5.2

Comparison of Output Voltage and THD among Different Resonant Converters.

		V_o(V)	Harmonic Statistics			Harmonic proportion (%)		
Converter	THD(%)		1st	3rd	5th	1st	3rd	5th
SRC	3.355	7.88	5.50	3.042	3.663	100	18.4	22.2
RPC	4.254	6.09	12.18	0.419	0.261	100	3.44	2.14
SRPC	4.341	15.9	59.87	2.485	0.625	100	4.15	1.04
CRD Γ-LC	2.407	12.0	23.29	0.517	0.179	100	2.22	0.77
V_i	—	—	18.83	6.366	3.820	100	33.8	18.5

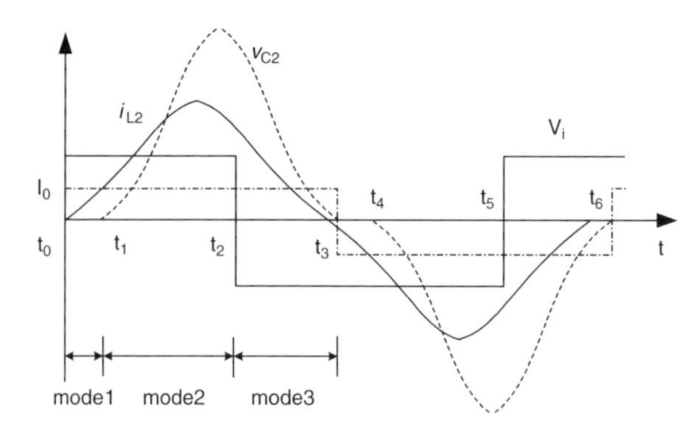

FIGURE 5.26
Switching waveforms for the discontinuous-mode cascade reverse double Γ-LC RPC.

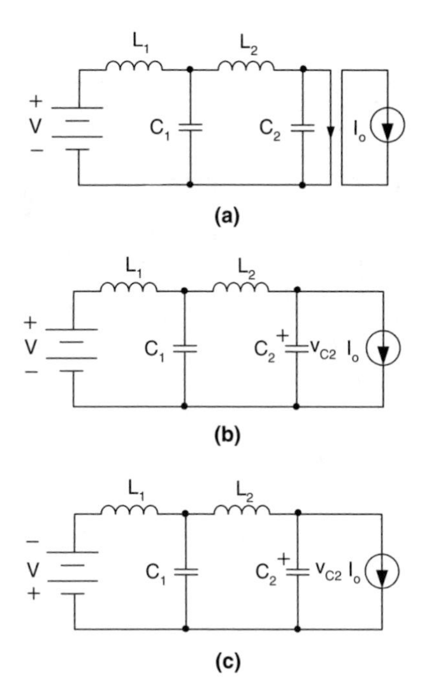

FIGURE 5.27
Discontinuous-mode equivalent circuits in half cycle.

$$A_5 = \begin{bmatrix} 0 & 0 & -1/L_1 & 0 & 0 & 0 \\ 0 & 0 & 1/L_2 & 0 & 0 & 0 \\ 1/C_1 & -1/C_2 & 0 & 0 & 0 & 0 \\ 0 & 0 & 0 & 0 & 0 & 0 \\ 0 & 0 & 0 & 0 & 0 & -1/L_f \\ 0 & 0 & 0 & 0 & 1/C_f & -1/RC_f \end{bmatrix}$$

$$B_5 = \begin{bmatrix} V/L_1 & 0 & 0 & 0 & 0 & 0 \end{bmatrix}^T$$

and

$$A_6 = A_5 \quad B_6 = \begin{bmatrix} -V/L_1 & 0 & 0 & 0 & 0 & 0 \end{bmatrix}^T$$

The augmented state-space equation in Section 5.4 is still valid besides that the state transition matrices describing two additional operating modes should be considered as well. The waveforms of simulation and experimental results under DCM operation are shown in Figure 5.28. Note that the

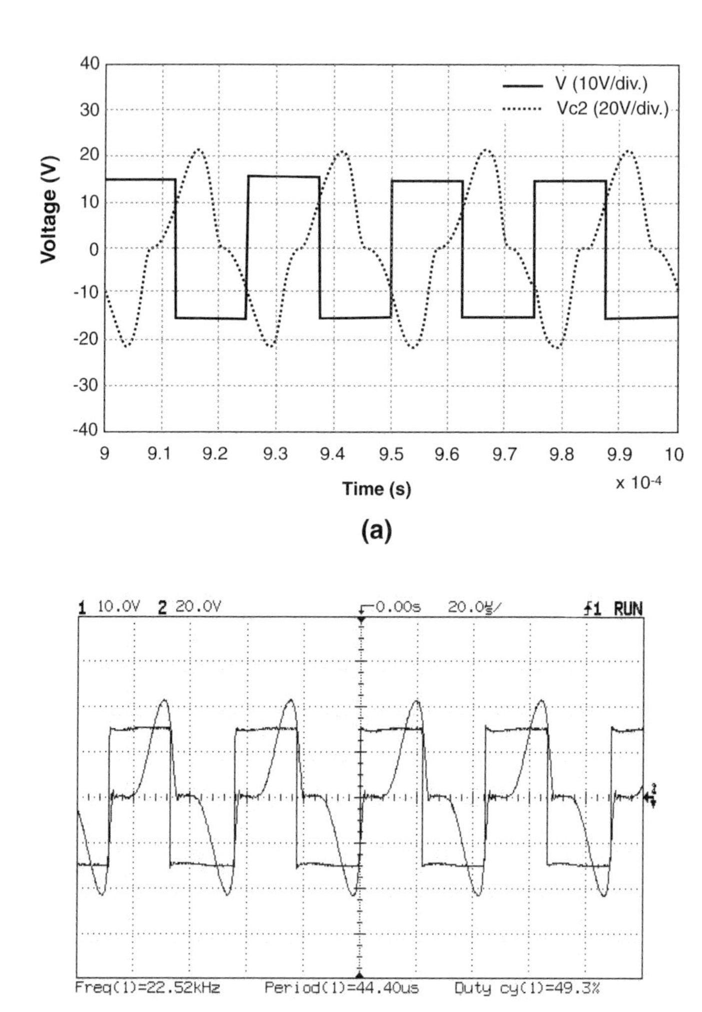

FIGURE 5.28
Simulation (a) and experimental (b) switching waveforms for DCM operation.

second capacitor voltage is discontinuous at some intervals, which verifies the description of the theoretical analysis.

In practice, the discontinuous conduction mode is often dependent on the switching frequency and the equivalent load current I_0. Namely, under a certain load current, when the switching frequency is increased, the operating state of the resonant converter will transfer from DCM to CCM. Conversely, if the switching frequency is invariant, the increase of the load current will lead to the occurrence of DCM. Thus, it is very significant to

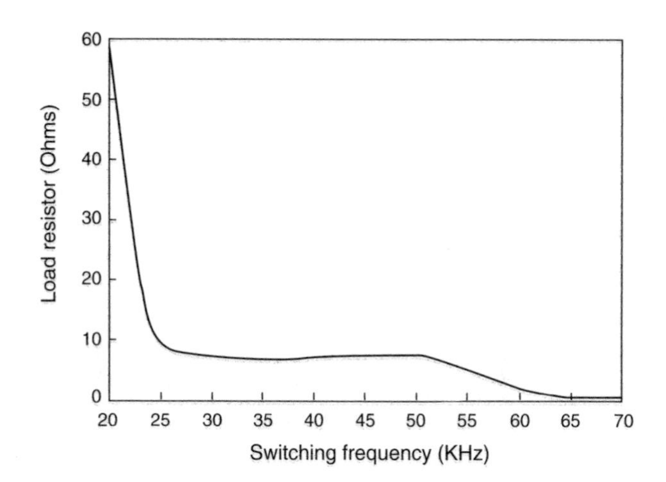

FIGURE 5.29
Critical load resistor vs. the switching frequency.

find the relationship between the load current and the switching frequency, in other words, to find the critical load current J_{cr} under a certain frequency.

In many open literatures, the critical load current is always obtained by solving the differential equations and identifying the respective duration of various subintervals. For instance, the normalized critical load current of RPC is given:

$$J_{cr} = \sqrt{\sin^2(\gamma / 2) + \sin^2(\gamma) / 4} - \sin(\gamma) / 2 \tag{5.71}$$

This conclusion is very simple and straightforward to describe the relationship between f_s and I_0. However, it is acquired at the cost of very complicated calculations and derivations. For SRPC, there is no similar analytical expression given in the literature. In fact, with the increase of the quantity of the resonant components, the order of the differential equations becomes very large so that it seems impossible to solve them by pen and paper. Once again, the numerical calculation offers a very useful solution to overcome such a problem.

The numerical analysis method starts at solving the higher order state equations using recursion algorithm. Once the steady state is achieved, it is up to the algorithm that judges whether the critical load current is found correctly. The drawbacks of this method are in two aspects: one is the discrete result; the other is time-consuming. However, the obtained results are proved to be precise and useful so long as the step of the loop is small enough. The critical load resistor and critical load current vs. the switching frequency are shown in Figure 5.29 and Figure 5.30 respectively.

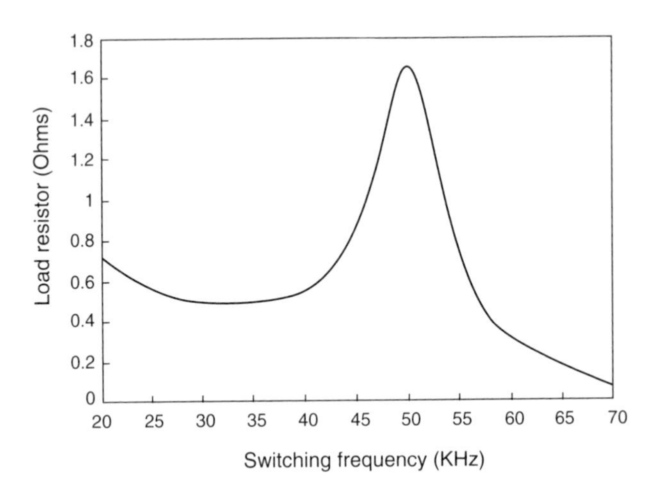

FIGURE 5.30
Critical load current vs. the switching frequency.

Bibliography

Agarwal, V. and Bhat, A.K.S., Small-signal analysis of the LCC-type parallel resonant converter using discrete time domain modeling, in *Proceedings of IEEE Power Electronics Specialists Conference*, 1994, p. 805.

Bhat, K.S., Analysis and design of a series-parallel resonant converter with capacitive output filter, *IEEE Transactions on Industry Applications*, 27, 523, 1991.

Bhat, K.S. and Swamy, M.M., Analysis and design of a parallel resonant converter including the effect of high-frequency transformer, in *Proceedings of IEEE Power Electronics Specialists Conference*, 1989, p. 768.

Deb, S., Joshi, A., and Doradla, S.R., A novel frequency domain model for a parallel resonant converter, *IEEE Transactions on Power Electronics*, 3, 208, 1988.

Forsyth, A. J., Ho, Y.K. E., and Ong, H.M., Comparison of small-signal modeling techniques for the series-parallel resonant converter, in *Proceedings of IEE Conference on Power Electronics and Variable Speed Drives*, 1994, p. 268.

Johnson, S.D., Steady-state analysis and design of the parallel resonant converter, *IEEE Transactions on Power Electronics*, 3, 93, 1988.

Kang, Y.G. and Upadhyay, A.K., Analysis and design of a half-bridge parallel resonant converter, *IEEE Transactions on Power Electronics*, 3, 254, 1988.

Kit, S.K., *Recent Developments in Resonant Power Conversion*, CA, Intertec Communications, 1988.

Liu, K.H. and Lee, F.C., Zero-voltage switching technique in DC/DC converters, in *Proceedings of IEEE Power Electronics Specialists Conference*, 1986, 58.

Middlebrook, R.D. and Cúk, S., A general unified approach to modeling switching converter power stages, in *Proceedings of IEEE Power Electronics Specialists Conference*, 1976, p. 18.

Nathan, B.S. and Ramanarayanan, V., Analysis, simulation and design of series resonant converter for high voltage applications, in *Proceedings of IEEE International Conference on Industrial Technology 2000*, 2000, p. 688.

Ranganathan, V.T., Ziogas, P.D., and Stefenovic, V.R., A regulated DC-DC voltage source converter using high-frequency link, in *Proceedings of IEEE Industry Application Society Conference*, 1981, p. 917.

Sanders, S., Noworolski, J., Liu, X., and Verghese, G., Generalized averaging method for power conversion circuits, in *Proceedings of IEEE Power Electronics Specialists Conference*, 1989, p. 273.

Swamy, M.M. and Bhat, A.K.S., A comparison of parallel resonant converters operating in lagging power factor mode, *IEEE Transactions on Power Electronics*, 9, 181, 1994.

Verghese, G., Elbuluk, M., and Kassakian, J., Sampled-data modeling for power electronics circuits, in *Proceedings of IEEE Power Electronics Specialists Conference*, 1984, p. 316.

Vorperian, V. and Cúk, S., A complete DC analysis of the series resonant converter, in *Proceedings of IEEE Power Electronics Specialist Conference*, 1982, p. 85.

Vorpérian, V. and Cúk, S., Small-signal analysis of resonant converters, in *Proceedings of IEEE Power Electronics Specialists Conference*, 1983, p. 269.

Yang, E.X., Choi, B.C., Lee, F.C., and Cho, B.H., Dynamic analysis and control design of LCC resonant converter, in *Proceedings of IEEE Power Electronics Specialists Conference*, 1992, p. 941.

Zhu, J.H. and Luo, F.L., Cascade reverse double Γ-LC resonant power converter, in *Proceedings of IEEE-PEDS'03*, Singapore, 2003, p. 326.

APPENDIX

Parameters Used in Small-Signal Modeling

$$k_v = \frac{4}{\pi}$$

$$k_s = \frac{2}{\pi}\frac{V_{2s}}{A_p}$$

$$k_c = \frac{2}{\pi}\frac{V_{2c}}{A_p}$$

$$Z_{L1} = \Omega_s L_1$$

$$Z_{L2} = \Omega_s L_2$$

$$G_s = \Omega_s C_1$$

$$E_{1s} = \omega_0 L_1 I_{1c}$$

$$E_{1c} = \omega_0 L_1 I_{1s}$$

$$E_{2s} = \omega_0 L_2 I_{2c}$$

$$E_{2c} = \omega_0 L_2 I_{2s}$$

$$J_{1s} = \omega_0 C_1 V_{1c}$$

$$J_{1c} = \omega_0 C_1 V_{1s}$$

$$J_{2s} = \omega_0 C_2 V_{2c}$$

$$J_{2c} = -\omega_0 C_2 V_{2s}$$

$$g_{ss} = \frac{g_e \alpha^2}{\alpha^2 + \beta^2}$$

$$g_{cc} = \frac{g_e \beta^2}{\alpha^2 + \beta^2}$$

$$g_{sc} = \Omega_s C_2 + \frac{\alpha\beta}{\alpha^2 + \beta^2}$$

$$g_{cs} = -\Omega_s C_2 + \frac{\alpha\beta}{\alpha^2 + \beta^2}$$

$$V_e = \frac{4}{\pi} V_g$$

$$g_e = \frac{8}{\pi^2 R}$$

$$A_p = \frac{1}{\sqrt{\alpha^2 + \beta^2}} V_e$$

$$\alpha = g_e \Omega_s L(\Omega_s^2 LC - 2)$$

$$\beta = \Omega_s^4 L^2 C^2 - 3\Omega_s^2 LC + 1$$

6

DC Energy Sources for DC/DC Converters

The DC/DC converter is used to convert a DC source voltage to another DC voltage actuator (user). In a DC/DC converter system the main parts are the DC voltage source, switches, diodes, inductors/capacitors, and load. This chapter introduces the various DC energy sources that are usually employed in DC/DC converters.

6.1 Introduction

In a DC/DC converter system the initial energy source is a DC voltage source with certain voltage and very low internal impedance, which can be usually omitted. This means the used DC voltage source is ideal. The DC voltage source can be a battery, a DC bus (usually equipped in factories and laboratories), a DC generator, and an AC/DC rectifier. As is well-known, the battery, DC bus, and DC generator can be considered an ideal voltage source. They will be not discussed in this book. AC/DC rectifiers are widely applied in industrial applications and research centers since it is easily constructed and less costly. In this chapter the AC/DC rectifiers will be discussed in detail. AC/DC rectifiers can be grouped as follows:

- Single-phase half-wave diode rectifier
- Single-phase full-wave bridge diode rectifier
- Three-phase bridge diode rectifier
- Single-phase half-wave thyrister rectifier
- Single-phase full-wave bridge thyrister rectifier
- Three-phase bridge thyrister rectifier
- Other devices rectifiers

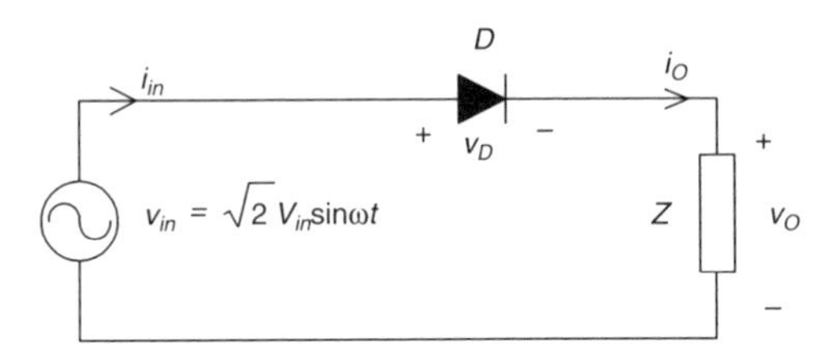

FIGURE 6.1
Half-wave diode rectifier.

6.2 Single-Phase Half-Wave Diode Rectifier

The single-phase half-wave diode rectifier is shown in Figure 6.1, it is the simplest rectifier circuit. The load Z in the figure can be any type such as resistor, inductor, capacitor, back electromotive force (EMF), and/or a combination of these. This rectifier can rectify the AC input voltage into DC output voltage. The analysis of the circuit is based on the assumption that a diode as an ideal component is used for the rectification. A diode forward biased will conduct without forward voltage-drop and resistance. A diode reverse biased will be blocked, and likely an open circuit. Since the used diode is not continuously conducted, the output current is always discontinuous in some part of negative half-cycle.

6.2.1 Resistive Load

A single-phase half wave rectifier with a purely resistive load (R) is shown in Figure 6.2a, and its input/output voltage v_{in} and v_O, and input/output current i_{in} and i_O waveforms are shown in Figure 6.2b and c. The AC supply voltage v_{in} is sinusoidal, the output voltage and current obey Ohm's law. Therefore the output voltage and current are sinusoidal half-waveforms

$$v_{in}(t) = \sqrt{2}V_{in}\sin\omega t \tag{6.1}$$

$$v_O(t) = \begin{cases} \sqrt{2}V_{in}\sin\omega t & 0 \le \omega t \le \pi \\ 0 & \pi < \omega t < 2\pi \end{cases} \tag{6.2}$$

$$i_{in}(t) = i_O(t) = \begin{cases} \dfrac{\sqrt{2}V_{in}}{R}\sin\omega t & 0 \le \omega t \le \pi \\ 0 & \pi < \omega t < 2\pi \end{cases} \tag{6.3}$$

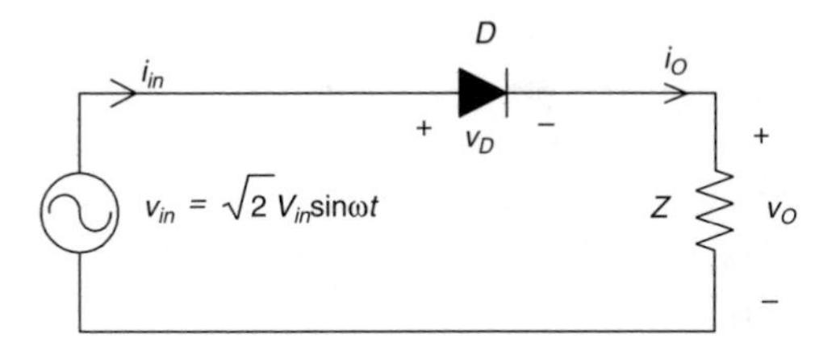

(a) Circuit diagram

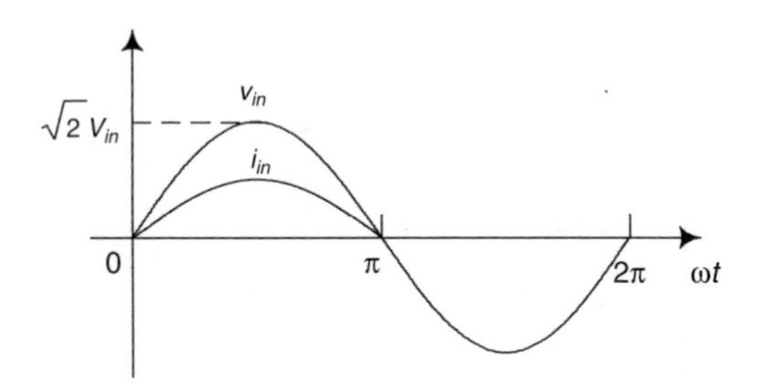

(b) Input voltage and current

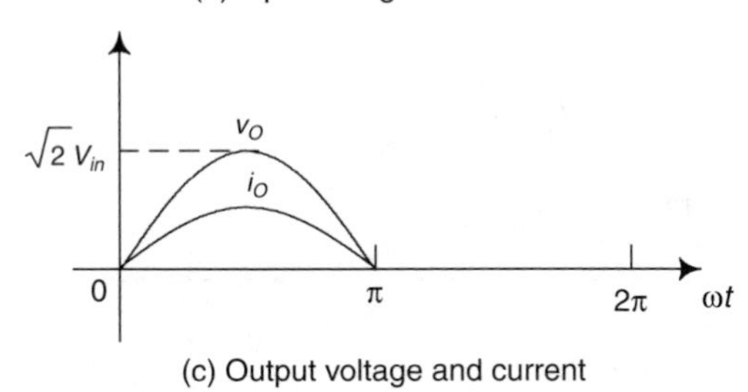

(c) Output voltage and current

FIGURE 6.2
A single-phase half wave rectifier with a purely resistive load (R).

Where V_{in} is the root-mean-square (RMS) value of the input voltage. The input wave is a sinusoidal waveform, the corresponding output is half-wave of a sinusoidal waveform for both voltage and current without angle shift between voltage and current. The output DC average voltage and current are

$$V_{O-av} = \frac{\sqrt{2}}{\pi}V_{in} = 0.45V_{in} \tag{6.4}$$

$$I_{O-av} = 0.45\frac{V_{in}}{R} \tag{6.5}$$

6.2.2 Inductive Load

A single-phase half wave rectifier with an inductive load (a resistor R plus an inductor L) is shown in Figure 6.3a. The input voltage and current v_{in} and i_{in} waveforms are shown in Figure 6.3b, output voltage v_O in c and output

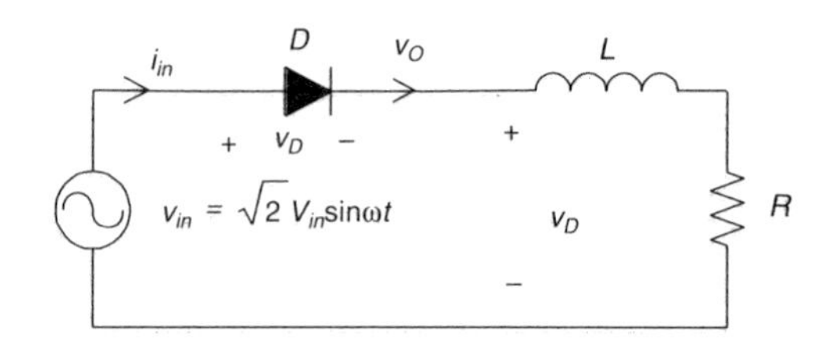

(a) Circuit diagram

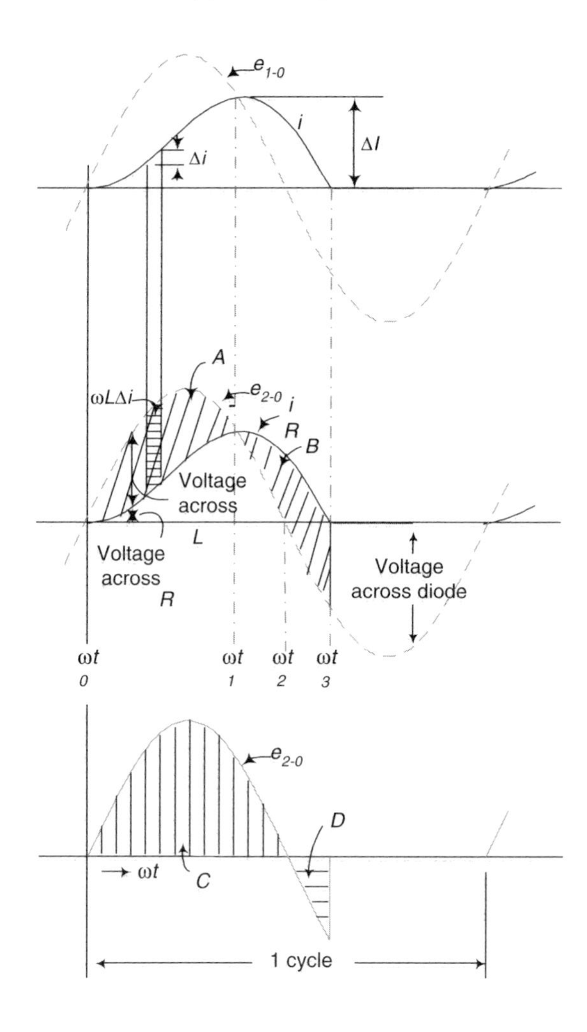

(b),(c), and (d)

FIGURE 6.3
A single-phase half wave rectifier with an inductive load ($R + L$).

current i_O in d. The AC supply voltage is sinusoidal, the output voltage and current obey Ohm's law. The impedance of load is

$$Z = R + j\omega L = \sqrt{R^2 + (\omega L)^2} \angle \phi$$

where

$$|Z| = \sqrt{R^2 + (\omega L)^2} \quad \text{and} \quad \phi = \tan^{-1}\frac{\omega L}{R}$$

Input voltage is

$$v_{in}(t) = \sqrt{2}V_{in}\sin\omega t \tag{6.1}$$

Output voltage is

$$v_O(t) = \begin{cases} \sqrt{2}V_{in}\sin\omega t & 0 \le \omega t \le (\pi + \phi) \\ 0 & (\pi + \phi) < \omega t < 2\pi \end{cases} \tag{6.6}$$

Where V_{in} is the RMS value of the input voltage. The input wave is sinusoidal waveform, the corresponding output is a partial sinusoidal waveform more than half-cycle. Since it is negative value in the negative half-cycle, the output DC average voltage is

$$V_{O-av} = \frac{\sqrt{2}}{\pi}(1 - \cos\phi)V_{in} = 0.45(1 - \cos\phi)V_{in} \tag{6.7}$$

The input and output current waveform is no longer a sinusoidal waveform.

$$i_{in}(t) = i_O(t) = \begin{cases} \frac{\sqrt{2}V_{in}}{|Z|}[\sin(\omega t - \phi) + \sin\phi \cdot e^{-t/\tau}] & 0 \le \omega t \le (\pi + \phi) \\ 0 & (\pi + \phi) < \omega t < 2\pi \end{cases} \tag{6.8}$$

where τ is the load time constant $\tau = L/R$. The input and output current average value is

$$I_{O-av} = 0.45\frac{V_{in}}{R}(1 - \cos\phi) \tag{6.9}$$

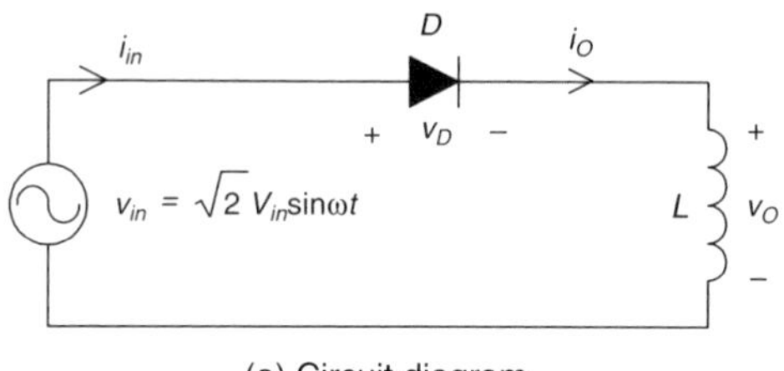

(a) Circuit diagram

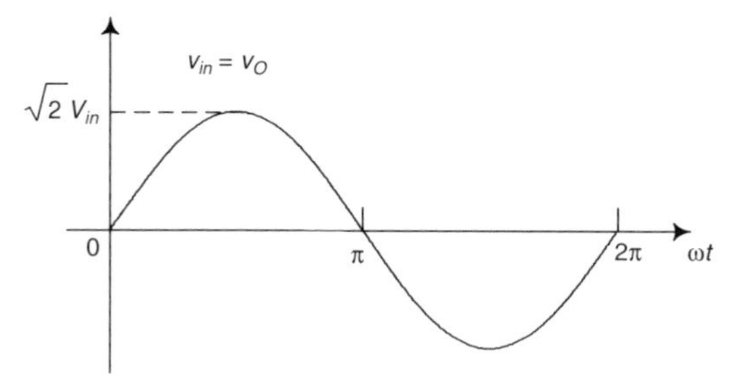

(b) Input and output voltages

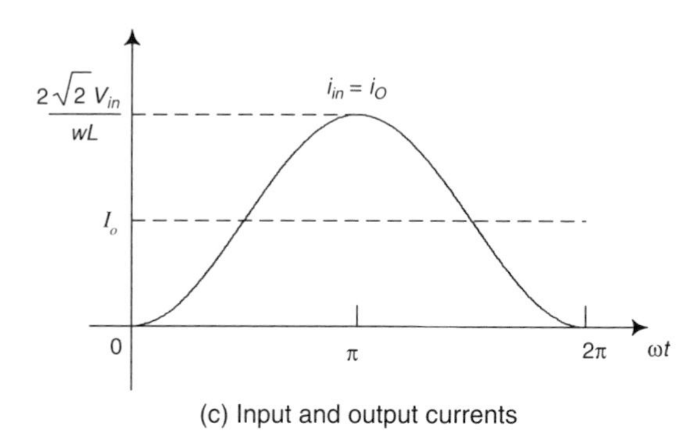

(c) Input and output currents

FIGURE 6.4
A single-phase half wave rectifier with a pure inductive load (L).

6.2.3 Pure Inductive Load

A single-phase half wave rectifier with a pure inductive load (an inductor L only) is shown in Figure 6.4a, and its input/output voltage v_{in} and v_O and input/output current i_{in} and i_O waveforms are shown in Figure 6.4b and c. The circuit will be analyzed for the relationship of the output to input. The AC supply voltage is sinusoidal, the output voltage can follow it in time. The impedance of load is

$$Z = j\omega L = \omega L \angle \pi / 2$$

Input and output voltage is

$$v_{in}(t) = v_O(t) = \sqrt{2}V_{in} \sin \omega t \tag{6.1a}$$

Where V_{in} is the RMS value of the input voltage. The input wave is sinusoidal waveform, the corresponding output can be a full sinusoidal waveform too. Since it is negative value in full negative half-cycle, the output DC average voltage is 0.

The input and output current waveform is a sinusoidal waveform.

$$i_{in}(t) = i_O(t) = \frac{\sqrt{2}V_{in}}{\omega L}(1 - \cos \omega t) \tag{6.10}$$

6.2.4 Back EMF Plus Resistor Load

A single-phase half wave rectifier with a resistor plus an EMF load (a resistor R plus an EMF) is shown in Figure 6.5a, and its input/output voltage v_{in} and v_O and input/output current i_{in} and i_O waveforms are shown in Figure 6.5b and c. The EMF value is E which is smaller than the input peak voltage $\sqrt{2}V_{in}$. Suppose an auxiliary parameter m is

$$m = E/\sqrt{2}V_{in} < 1$$

and

$$\alpha = \sin^{-1} m = \sin^{-1} \frac{E}{\sqrt{2}V_{in}}$$

The circuit will be analyzed for the relationship of the output to input. The AC supply voltage is sinusoidal, the output voltage and current obey Ohm's law. The impedance of load is

Input voltage is
$$v_{in}(t) = \sqrt{2}V_{in} \sin \omega t \tag{6.1}$$

Input voltage is
$$v_O(t) = \begin{cases} \sqrt{2}V_{in} \sin \omega t & \alpha \le \omega t \le (\pi - \alpha) \\ E & (\pi - \alpha) < \omega t < (2\pi + \alpha) \end{cases} \tag{6.11}$$

Where V_{in} is the RMS value of the input voltage. The input wave is a sinusoidal waveform, the corresponding output is a partial sinusoidal waveform less than half-cycle. The output DC average voltage is

$$V_{O-av} = \frac{\sqrt{2}V_{in}}{\pi} \cos \alpha + E(\frac{1}{2} + \frac{\alpha}{\pi}) > E \tag{6.12}$$

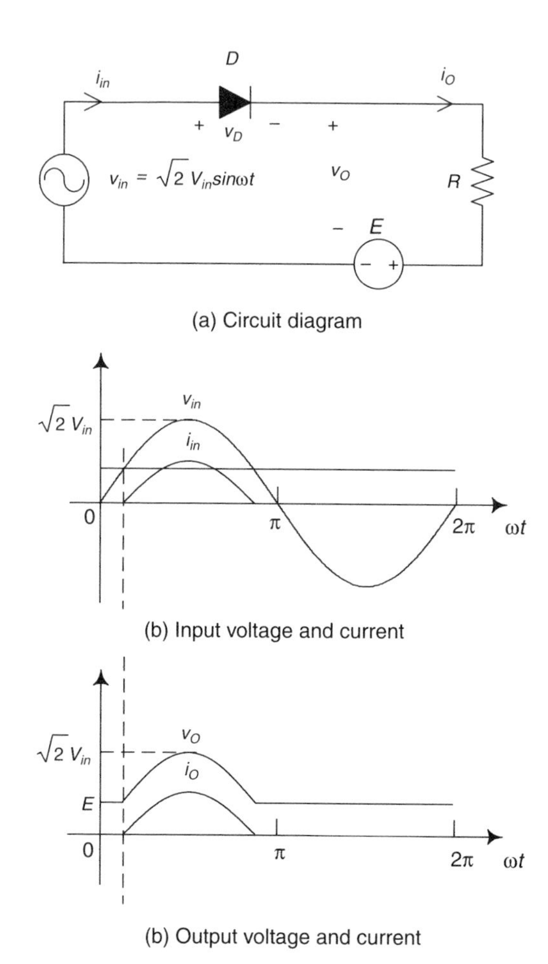

FIGURE 6.5
Single-phase half wave rectifier with an EMF plus resistor (R + EMF).

The input and output current waveform is no longer a sinusoidal waveform.

$$i_{in}(t) = i_O(t) = \begin{cases} \frac{1}{R}(\sqrt{2}V_{in}\sin\omega t - E) & \alpha \le \omega t \le (\pi - \alpha) \\ 0 & (\pi - \alpha) < \omega t < (2\pi + \alpha) \end{cases} \tag{6.13}$$

The input and output current average value is

$$I_{O-av} = \frac{0.45V_{in}}{2R}[2\cos\alpha - m(\pi - 2\alpha)] \tag{6.14}$$

6.2.5 Back EMF Plus Inductor Load

A single-phase half wave rectifier with an EMF and inductive load (an EMF plus an inductor L) is shown in Figure 6.6a, and its input/output voltage v_{in} and v_O and input/output current i_{in} and i_O waveforms are shown in Figure 6.6b and c. The circuit will be analyzed for the relationship of the output to input. The AC supply voltage is sinusoidal waveform:

$$v_{in}(t) = \sqrt{2}V_{in} \sin \omega t \tag{6.1}$$

Output voltage is $$v_O(t) = \begin{cases} \sqrt{2}V_{in} \sin \omega t & \alpha \le \omega t \le (\pi + \gamma) \\ E & (\pi + \gamma) < \omega t < (2\pi + \alpha) \end{cases} \tag{6.15}$$

Where $$\int_{\alpha}^{\pi-\alpha} (\sqrt{2}V_{in} \sin \omega t - E) d(\omega t) = \int_{\pi-\alpha}^{\pi-\alpha+\gamma} (E - \sqrt{2}V_{in} \sin \omega t) d(\omega t)$$

Where v_{in} is the RMS value of the input voltage. The input wave is sinusoidal waveform, the corresponding output is a partial sinusoidal waveform. The output DC average voltage is E.

The input and output current waveform is no longer a sinusoidal waveform.

$$i_{in}(t) = i_O(t)$$

$$= \begin{cases} \dfrac{\sqrt{2}V_{in}}{\omega L}[(\cos \alpha - \cos \omega t) - m(\omega t - \alpha)] & \alpha \le \omega t \le (\pi + \alpha + \gamma) \\ 0 & (\pi + \alpha + \gamma) < \omega t < (2\pi + \alpha) \end{cases} \tag{6.16}$$

The input and output current average value is

$$I_{O-av} = \frac{0.45V_{in}}{2\omega L}[\gamma \cos \gamma - \sin(\alpha + \gamma) + \sin \alpha - \frac{m}{2}\alpha^2] \tag{6.17}$$

6.3 Single-Phase Bridge Diode Rectifier

Single-phase full-wave diode rectifier has two forms:

- Bridge (Graetz)
- Center-tap (mid-point)

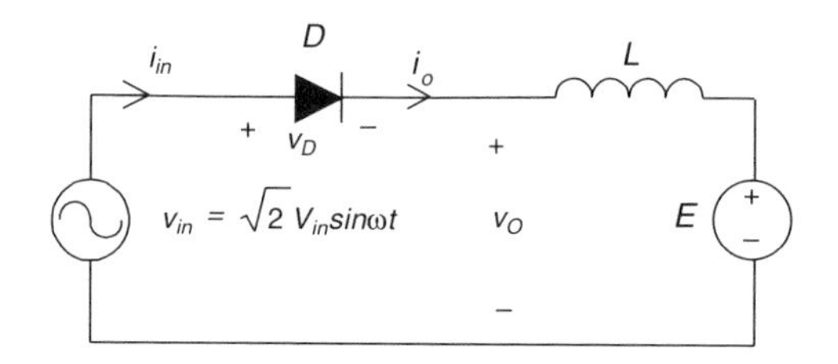

(a) Circuit diagram

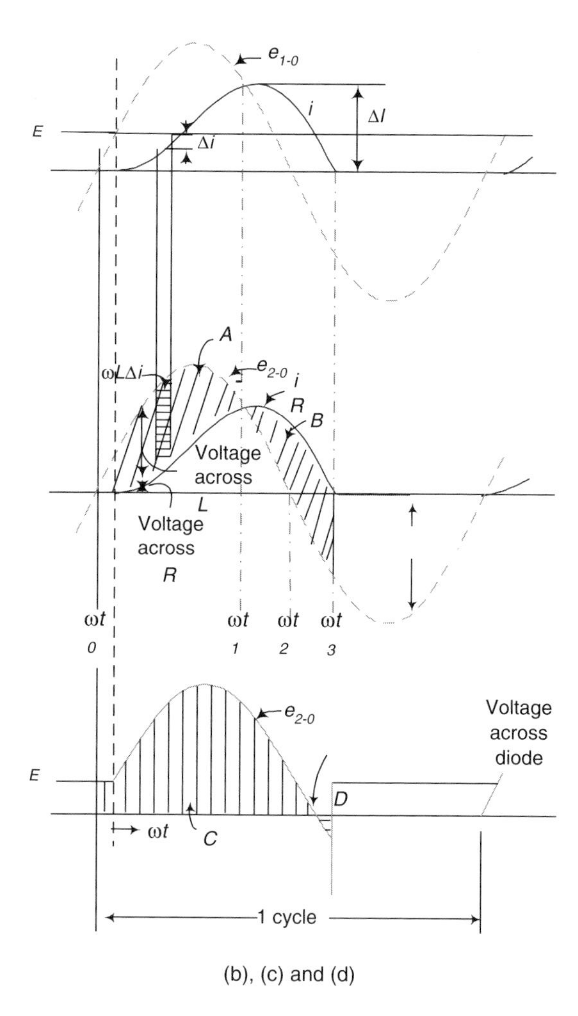

(b), (c) and (d)

FIGURE 6.6
Single-phase half wave rectifier with EMF plus inductor (*EMF* + *L*).

These are shown in Figure 6.7a and b. The input and output waveforms are same in both circuits. Use Graetz type for the description in the following sections.

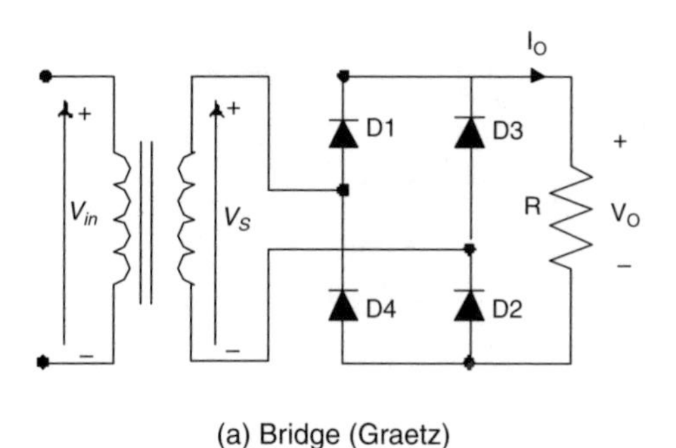

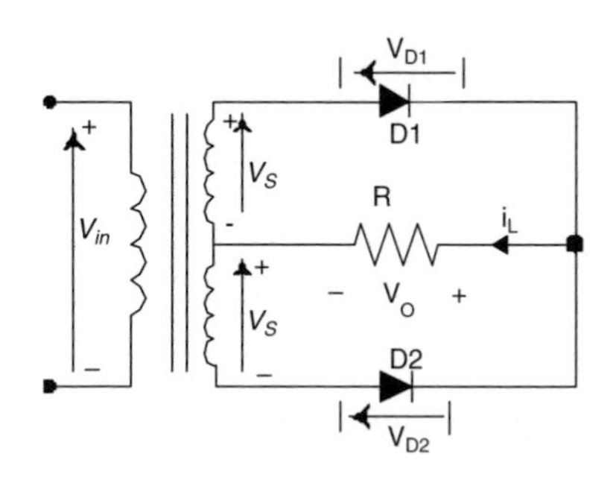

FIGURE 6.7
Single-phase full-wave diode rectifier.

6.3.1 Resistive Load

A single-phase full-wave diode rectifier with a purely resistive load is shown in Figure 6.8a, and its input/output voltage v_{in} and v_O and input/output current i_{in} and i_O waveforms are shown in Figure 6.8b and c. The circuit will be analyzed for the relationship of the output to input. The AC supply voltage is sinusoidal, the output voltage and current obey Ohm's law. Therefore the output voltage and current are sinusoidal half-waveforms.

$$v_{in}(t) = \sqrt{2}V_{in} \sin \omega t \tag{6.18}$$

$$v_O(t) = \begin{cases} \sqrt{2}V_{in} \sin \omega t & 0 \le \omega t \le \pi \\ \sqrt{2}V_{in} \sin(\omega t - \pi) & \pi < \omega t < 2\pi \end{cases} \tag{6.19}$$

$$i_{in}(t) = i_O(t) = \begin{cases} \dfrac{\sqrt{2}V_{in}}{R} \sin \omega t & 0 \le \omega t \le \pi \\ \dfrac{\sqrt{2}V_{in}}{R} \sin(\omega t - \pi) & \pi < \omega t < 2\pi \end{cases} \tag{6.20}$$

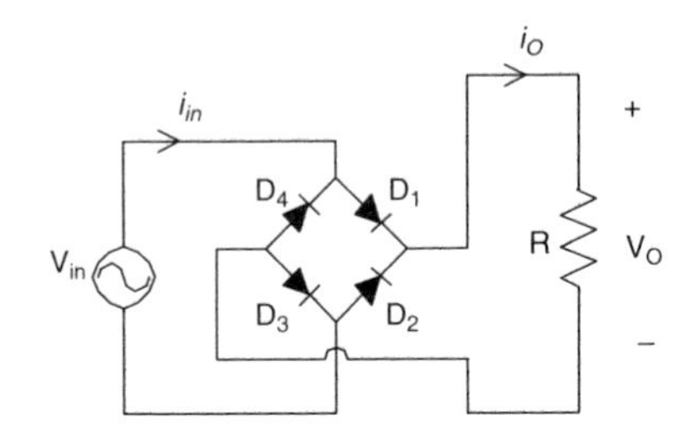

(a) Circuit diagram

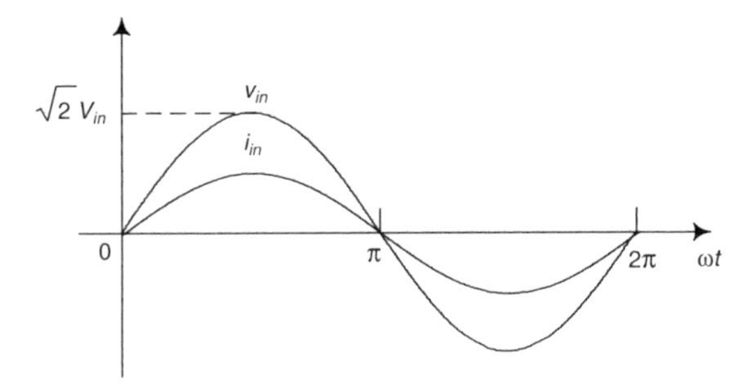

(b) Input voltage and current

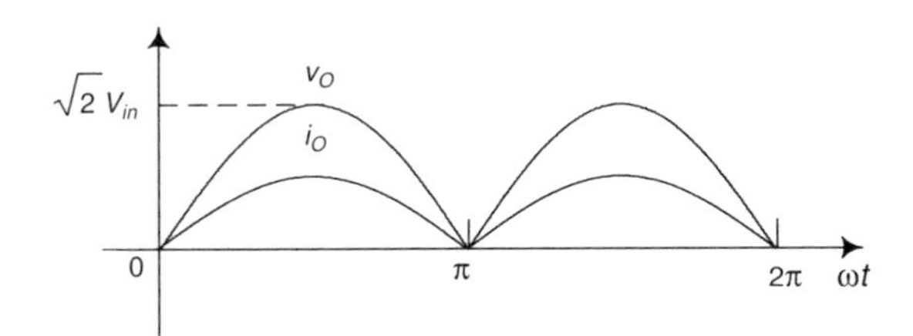

(c) Output voltage and current

FIGURE 6.8
Single-phase full-wave diode rectifier with a purely resistive load.

Where V_{in} is the RMS value of the input voltage. The input wave is sinusoidal waveform, the corresponding output is repeating half-wave sinusoidal waveform for both voltage and current without angle shift between voltage and current. The output is a DC voltage with ripple in the repeating frequency 2ω. After FFT analysis of the rectified waveform, harmonics components are shown in the frequency spectrum. From the spectrum, there are only nth ($n = 2k$) harmonics exsisting. The parameter ripple factor RF is defined

$$RF = \frac{V_{ac}}{V_{dc}} = \frac{\sqrt{\sum_{n=1}^{\infty} V_n}}{V_{dc}} \tag{6.21}$$

where V_{dc} is the DC component of the output voltage, which is the average value, V_n is the nth order harmonic component of the output voltage. The output DC average voltage and current are

$$V_{O-av} = \frac{2\sqrt{2}}{\pi} V_{in} = 0.9\ V_{in} \tag{6.22}$$

$$I_{O-av} = 0.9 \frac{V_{in}}{R} \tag{6.23}$$

6.3.2 Back EMF Load

A single-phase full-wave diode rectifier with an EMF plus resistor load (a resistor R plus an EMF) is shown in Figure 6.9a, and its input/output voltage v_{in} and v_O and input/output current i_{in} and i_O waveforms are shown in Figure 6.9b and c. The EMF value is E which is smaller than the input peak voltage $\sqrt{2}V_{in}$. Suppose an auxiliary parameter m:

$$m = E/\sqrt{2}V_{in} < 1$$

and

$$\alpha = \sin^{-1} m = \sin^{-1} \frac{E}{\sqrt{2}V_{in}}$$

The circuit will be analyzed for the relationship of the output to input. The AC supply voltage is sinusoidal, the output voltage and current obey Ohm's law. The impedance of load is

Input voltage is $$v_{in}(t) = \sqrt{2}V_{in} \sin \omega t \tag{6.1}$$

Output voltage is $$v_O(t) = \begin{cases} \sqrt{2}V_{in} \sin \omega t & \alpha \le \omega t \le (\pi - \alpha) \\ E & (\pi - \alpha) < \omega t < (\pi + \alpha) \end{cases} \tag{6.24}$$

Where V_{in} is the RMS value of the input voltage. The input wave is sinusoidal waveform, the corresponding output is a partial sinusoidal waveform less than half-cycle with repeating frequency of 2ω. The output DC average voltage is

$$V_{O-av} = \frac{2\sqrt{2}V_{in}}{\pi} \cos \alpha + \frac{2\alpha}{\pi} E > E \tag{6.25}$$

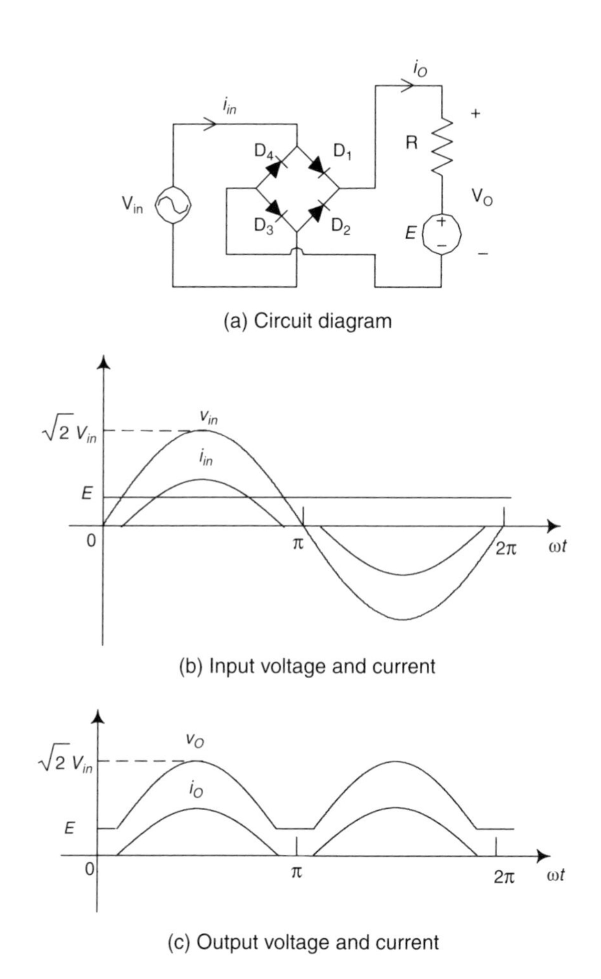

FIGURE 6.9
Single-phase full-wave *D*-rectifier with EMF plus resistor (*R* + *EMF*).

The input and output current waveform is no longer a sinusoidal waveform.

$$i_{in}(t) = i_O(t) = \begin{cases} \dfrac{1}{R}(\sqrt{2}V_{in}\sin\omega t - E) & \alpha \le \omega t \le (\pi - \alpha) \\ 0 & (\pi - \alpha) < \omega t < (\pi + \alpha) \end{cases} \tag{6.26}$$

The input and output current average value is

$$I_{O-av} = \frac{0.45V_{in}}{R}[\cos\alpha - m(\pi - 2\alpha)] \tag{6.27}$$

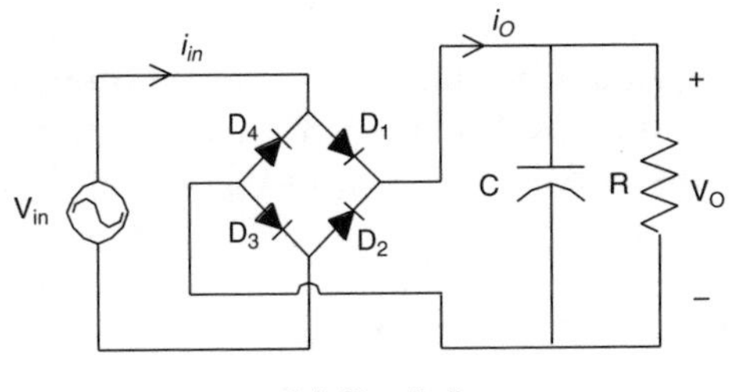

(a) Circuit diagram

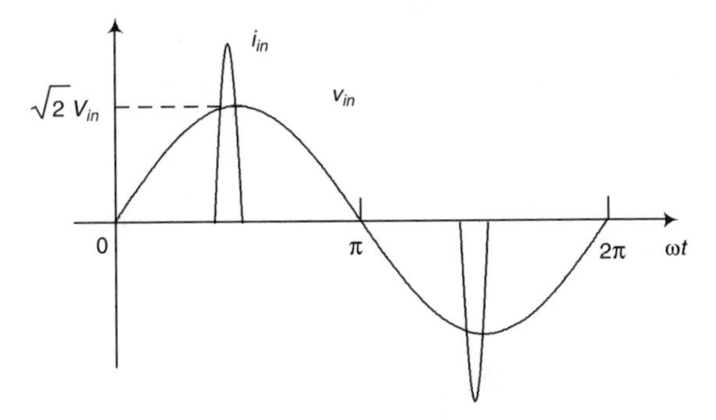

(b) Input voltage and current

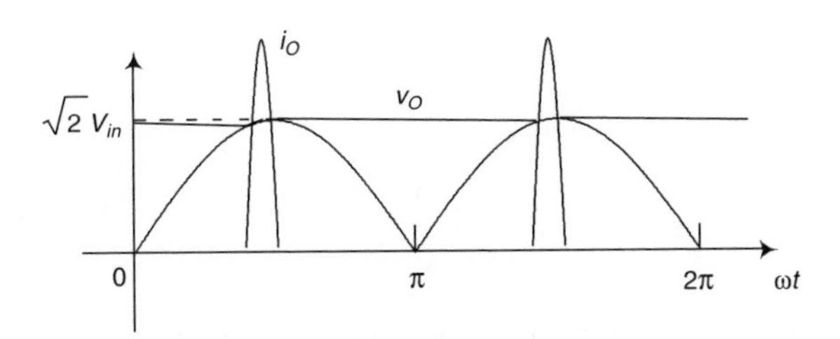

(c) Output voltage and current

FIGURE 6.10
Single-phase full wave *D*-rectifier with an capacitive load (*R* + C).

6.3.3 Capacitive Load

A single-phase full wave diode rectifier with a capacitive load (a resistor R plus a capacitor C) is shown in Figure 6.10a, and its input/output voltage v_{in} and v_O and input/output current i_{in} and i_O waveforms are shown in Figure 6.10b and c. The circuit will be analyzed for the relationship of the output to input. The AC supply voltage is sinusoidal. The load time constant $\tau = RC$.

Input voltage is
$$v_{in}(t) = \sqrt{2}V_{in} \sin \omega t \tag{6.1}$$

From previous section analysis, the peak value of output voltage is $V_{max} = \sqrt{2}V_{in}$. The minimum output DC voltage is V_{min}. The capacitor charges from V_{min} to V_{max} during t_1, discharges from V_{max} to V_{min} during $t_2 = T/2 - t_1 \cong 1/2f$ since usually $t_1 << t_2$.

The minimum output DC voltage is V_{min}

$$V_{min} = \sqrt{2}V_{in}e^{-\frac{t_2}{\tau}} \approx \sqrt{2}V_{in}e^{-\frac{1}{2fRC}} \tag{6.28}$$

Considering the peak-to-peak variation of the output DC voltage is

$$V_{pp} = \sqrt{2}V_{in}(1 - e^{-t_2/\tau}) \approx \sqrt{2}V_{in}(t_2 / \tau) = \frac{\sqrt{2}V_{in}}{2fRC} \tag{6.29}$$

Defining

$$\beta = \cos^{-1}\frac{\sqrt{2}V_{in} - V_{pp}}{\sqrt{2}V_{in}} = \cos^{-1}(e^{-t_2/\tau}) \approx \cos^{-1}(1 - \frac{1}{2fRC}) \tag{6.30}$$

Output voltage is

$$v_O(t) = \begin{cases} \sqrt{2}V_{in} \sin \omega t & (\frac{\pi}{2} - \beta) \leq \omega t \leq \pi / 2 \\ \sqrt{2}V_{in} - \frac{\omega t + \pi / 2}{\pi} V_{pp} & \frac{\pi}{2} < \omega t < (\frac{3\pi}{2} - \beta) \end{cases} \tag{6.31}$$

Where V_{in} is the RMS value of the input voltage. The input wave is sinusoidal waveform, the corresponding output is a partial sinusoidal waveform less than half-cycle. The repeating frequency is 2ω. The output DC average voltage is

$$V_{O-av} = \sqrt{2}V_{in}(1 - \frac{1}{4fRC}) \tag{6.32}$$

The capacitor charging and discharging current waveform is nearly a rectangular-wave.

$$v_C(t) = \begin{cases} \frac{V_{av}}{R}\frac{\pi - \beta}{\beta} & (\frac{\pi}{2} - \beta) \leq \omega t \leq \pi / 2 \\ -\frac{V_{av}}{R} & \frac{\pi}{2} < \omega t < (\frac{3\pi}{2} - \beta) \end{cases} \tag{6.33}$$

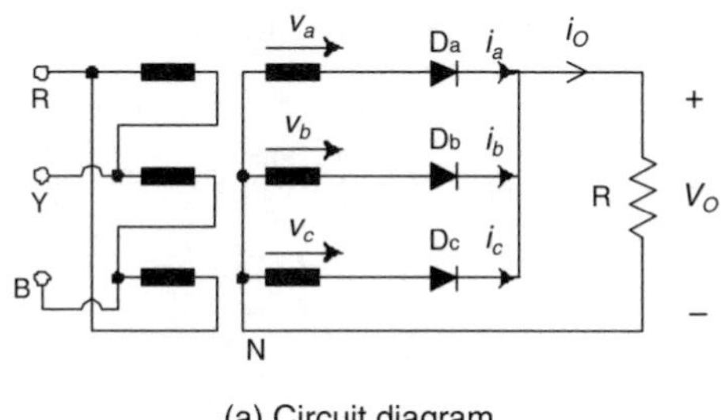

(a) Circuit diagram

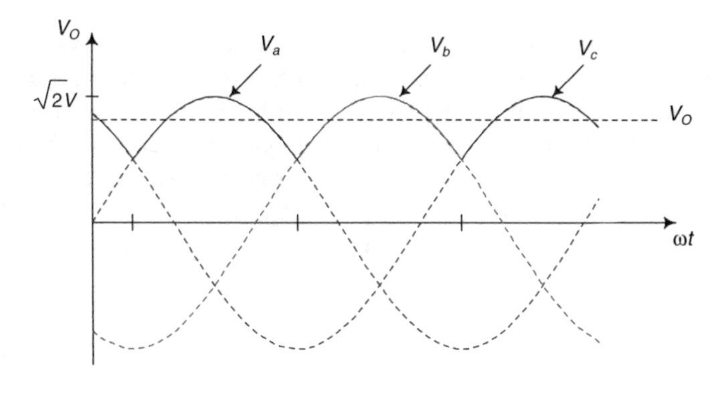

(b) Input voltage waveform

FIGURE 6.11
A three-phase half-bridge diode rectifier with a resistor (R).

6.4 Three-Phase Half-Bridge Diode Rectifier

A three-phase half-bridge diode rectifier is shown in Figure 6.11. The three phase AC suppliers have the same RMS value V_{in} with amplitude $\sqrt{2}V_{in}$ and frequency ω, the phase angle shift is 120°.

6.4.1 Resistive Load

A three-phase half-bridge diode rectifier with a purely resistive load is shown in Figure 6.11a, and its input/output voltage v_{in} and v_O and input/output current i_{in} and i_O waveforms are shown in Figure 6.11b. The circuit will be analyzed for the relationship of the output to input. The AC supply voltage is sinusoidal. Therefore the output voltage and current are sinusoidal half-waveforms.

$$v_{in}(t) = \sqrt{2}V_{in} \sin \omega t \tag{6.34}$$

$$v_O(t) = \sqrt{2}V_{in} \sin \omega t \qquad (\frac{2n\pi}{3} + \frac{\pi}{6}) \le \omega t \le (\frac{2n\pi}{3} + \frac{5\pi}{6}) \tag{6.35}$$

where $n = 1, 2, 3, \ldots$

$$i_{in}(t) = i_O(t) = \frac{\sqrt{2}V_{in}}{R} \sin \omega t \qquad (\frac{2n\pi}{3} + \frac{\pi}{6}) \le \omega t \le (\frac{2n\pi}{3} + \frac{5\pi}{6}) \tag{6.36}$$

Where V_{in} is the RMS value of the input voltage. The input wave is a sinusoidal waveform, the corresponding output is a repeating partial sinusoidal waveform for both voltage and current without angle shift between voltage and current. The output is a DC voltage with ripple in the repeating frequency 3ω. After FFT analysis of the rectified waveform, harmonic components are shown in the frequency spectrum. From the spectrum, there are only nth ($n = 3k$) harmonics existing. The parameter ripple factor RF is defined

$$RF = \frac{V_{ac}}{V_{dc}} = \frac{\sqrt{\sum_{n=1}^{\infty} V_n}}{V_{dc}} \tag{6.37}$$

where V_{dc} is the DC component of the output voltage which is the average value, V_n is the nth order harmonic component of the output voltage. The output DC average voltage and current are

$$V_{O-av} = \frac{3\sqrt{6}}{2\pi} V_{in} = 1.17 V_{in} \tag{6.38}$$

$$I_{O-av} = 1.17 \frac{V_{in}}{R} \tag{6.39}$$

6.4.2 Back EMF Load ($0.5\sqrt{2}V_{in} < E < \sqrt{2}V_{in}$)

A three-phase half-wave diode rectifier with an EMF plus resistor load (a resistor R plus an EMF) is shown in Figure 6.12. This section discusses the case that EMF value E is in condition: $0.5\sqrt{2}V_{in} < E < \sqrt{2}V_{in}$. Suppose an auxiliary parameter m:

$$0.5 < m = E/\sqrt{2}V_{in} < 1$$

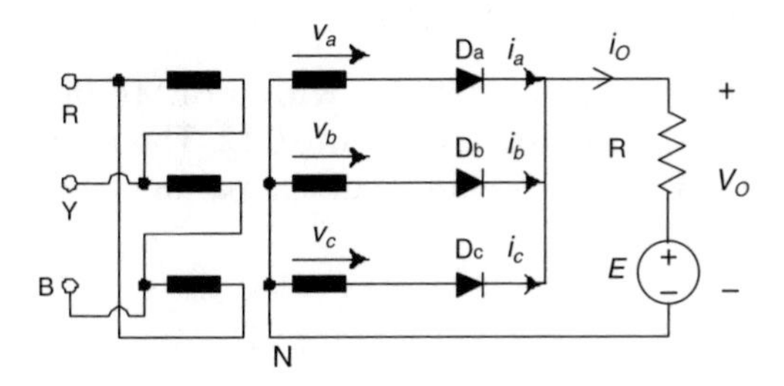

FIGURE 6.12
Three-phase half-wave diode rectifier with EMF plus resistor (R + EMF).

and

$$30° < \alpha = \sin^{-1} m = \sin^{-1} \frac{E}{\sqrt{2}V_{in}} < 90°$$

The circuit will be analyzed for the relationship of the output to input. The AC supply voltage is sinusoidal, the output voltage and current obey Ohm's law. The impedance of load is R

Input voltage is $$v_{in}(t) = \sqrt{2}V_{in} \sin \omega t \tag{6.1}$$

Output voltage is

$$v_O(t) = \begin{cases} \sqrt{2}V_{in} \sin \omega t & (\frac{2n\pi}{3} + \alpha) \leq \omega t \leq (\frac{2n\pi}{3} + \pi - \alpha) \\ E & other \end{cases} \tag{6.40}$$

Where V_{in} is the RMS value of the input voltage. The input wave is sinusoidal waveform, the corresponding output is a partial sinusoidal waveform less than half-cycle with repeating frequency of 3ω. The output DC average voltage is

$$V_{O-av} = \frac{3\sqrt{2}V_{in}}{\pi} \cos \alpha + (\frac{3\alpha}{\pi} - \frac{1}{2})E > E \tag{6.41}$$

The input and output current waveform is no longer a sinusoidal waveform.

$$i_{in}(t) = i_O(t) = \begin{cases} \frac{1}{R}(\sqrt{2}V_{in} \sin \omega t - E) & \alpha \leq \omega t \leq (\pi - \alpha) \\ 0 & (\pi - \alpha) < \omega t < (\pi + \alpha) \end{cases} \tag{6.42}$$

6.4.3 Back EMF Load ($E < 0.5\sqrt{2}V_{in}$)

A three-phase half-wave diode rectifier with an EMF plus resistor load (a resistor R plus an EMF) is shown in Figure 6.12. The EMF value E is in the condition $E < 0.5\sqrt{2}V_{in}$. Suppose an auxiliary parameter m:

$$m = E/\sqrt{2}V_{in} < 0.5$$

$$v_{in}(t) = \sqrt{2}V_{in}\sin\omega t \tag{6.34}$$

$$v_O(t) = \sqrt{2}V_{in}\sin\omega t \qquad (\frac{2n\pi}{3}+\frac{\pi}{6}) \le \omega t \le (\frac{2n\pi}{3}+\frac{5\pi}{6}) \tag{6.35}$$

where $n = 1, 2, 3, \ldots$

$$i_{in}(t) = i_O(t) = \begin{cases} \frac{1}{R}(\sqrt{2}V_{in}\sin\omega t - E) & (\frac{2n\pi}{3}+\frac{\pi}{6}) \le \omega t \le (\frac{2n\pi}{3}+\frac{5\pi}{6}) \\ 0 & other \end{cases} \tag{6.43}$$

Where V_{in} is the RMS value of the input voltage. The input wave is sinusoidal waveform, the corresponding output is repeating partial sinusoidal waveform for both voltage and current without angle shift between voltage and current. The output is a DC voltage with ripple in the repeating frequency 3ω. After FFT analysis of the rectified waveform, harmonic components are shown in the frequency spectrum. From the spectrum, there are only nth ($n = 3k$) harmonics existing.

Where V_{dc} is the DC component of the output voltage, which is the average value, V_n is the nth order harmonic component of the output voltage. The output DC average voltage and current are

$$V_{O-av} = \frac{3\sqrt{6}}{2\pi}V_{in} = 1.17V_{in} \tag{6.44}$$

6.5 Three-Phase Full-Bridge Diode Rectifier with Resistive Load

A three-phase full-bridge diode rectifier with a purely resistive load is shown in Figure 6.13a, and its input/output voltage v_{in} and v_O and input/output current i_{in} and i_O waveforms are shown in Figure 6.13b. The three phase AC suppliers have same RMS value V_{in} with amplitude $\sqrt{2}V_{in}$ and frequency ω, and phase angle shift of 120° each other. The circuit will be analyzed for the

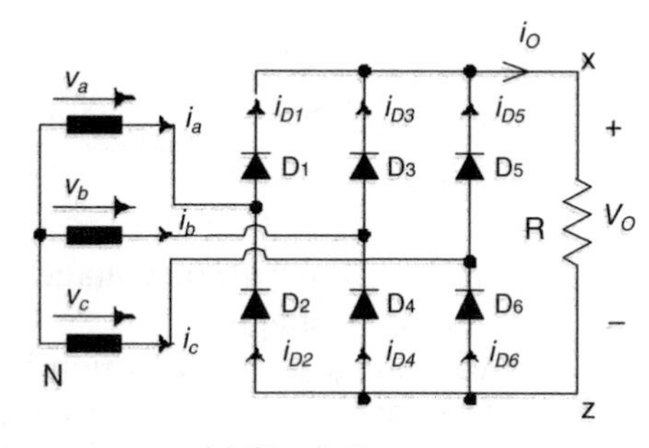

(a) Circuit diagram

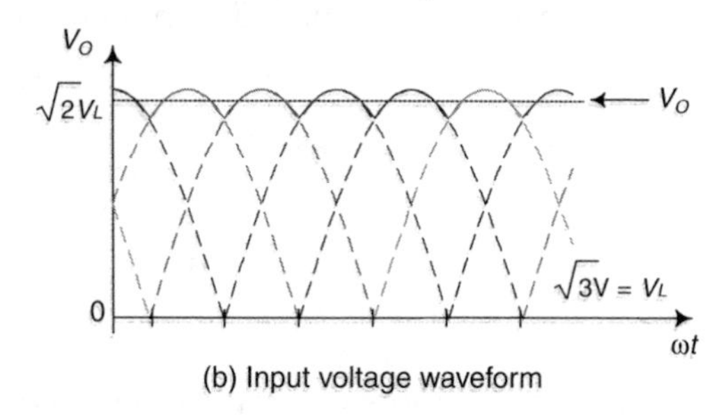

(b) Input voltage waveform

FIGURE 6.13
Three-phase full-bridge diode rectifier with a resistor (R).

relationship of the output to input. The AC supply voltage is sinusoidal. Therefore the output voltage and current are sinusoidal half-waveforms

$$v_{in}(t) = \sqrt{2}V_{in} \sin \omega t \tag{6.34}$$

$$v_O(t) = \sqrt{6}V_{in} \sin \omega t \qquad \frac{n\pi}{3} \le \omega t \le \frac{(n+1)\pi}{3} \tag{6.45}$$

where $n = 1, 2, 3, \ldots$

$$i_{in}(t) = i_O(t) = \frac{\sqrt{6}V_{in}}{R} \sin \omega t \qquad \frac{n\pi}{3} \le \omega t \le \frac{(n+1)\pi}{3} \tag{6.46}$$

Where V_{in} is the RMS value of the input voltage. The input wave is a sinusoidal waveform, the corresponding output is a repeating partial sinusoidal waveform for both voltage and current without angle shift between voltage and current. The output is a DC voltage with ripple in the repeating frequency 6 ω. After FFT analysis of the rectified waveform, harmonic components are shown in the frequency spectrum. From the spectrum, there are only nth ($n = 6k$) harmonics existing. The parameter ripple factor RF is defined

$$RF = \frac{V_{ac}}{V_{dc}} = \frac{\sqrt{\sum_{n=1}^{\infty} V_n}}{V_{dc}} = 0.054 \tag{6.47}$$

where V_{dc} is the DC component of the output voltage, V_n is the nth order harmonic component of the output voltage.

The input voltage is an AC voltage with distortion in the repeating frequency 6 ω. After FFT analysis of the supplying waveform, the harmonic components are shown in the frequency spectrum. From the spectrum, there are only nth ($n = 6k \pm 1$) harmonics existing. The parameter total harmonic distortion (THD) is defined

$$THD = \frac{V_{ac}}{V_{fund}} = \frac{\sqrt{\sum_{n=2}^{\infty} V_n}}{V_{fund}} = 0.046 \tag{6.48}$$

where V_{fund} is the fundamental component of the input voltage, V_n is the nth order harmonic component of the input voltage.

The output DC average voltage and current are

$$V_{O-av} = \frac{3\sqrt{6}}{\pi} V_{in} = 2.34\, V_{in} \tag{6.49}$$

$$I_{O-av} = 2.34 \frac{V_{in}}{R} \tag{6.50}$$

Since the output DC voltage ripple is very small ($RF = 0.054$), the EMF load usually has the condition $E < \sqrt{6}\, V_{in}$. There is no need to spend time discussing this case.

6.6 Thyristor Rectifiers

A thyristor is a silicon-controlled rectifier (SCR). It is a four-layer p-n-p-n semiconductor device forming three junctions J1-J2-J3, as shown in Figure 6.14a. It has three external electrodes: anode, cathode, and gate. This structure can be considered as a two-type transistor in a cascade connection shown in Figure 6.14b and c. Its characteristics will be discussed in the next section. It is controlled by a firing pulse with shifting firing angle (α). When α = 0° the characteristics of a thyristor is the same as those of a diode.

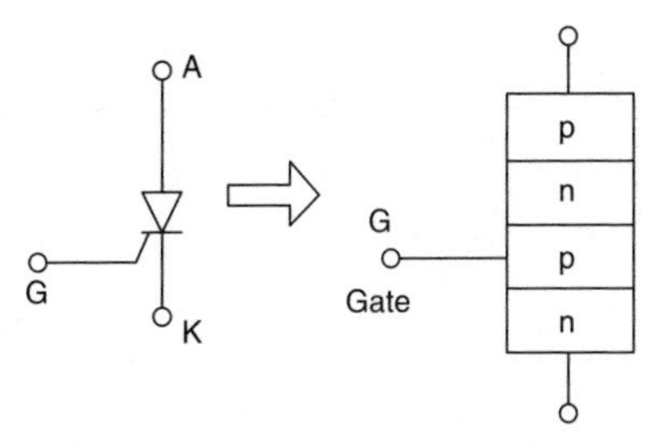

(a) Thyrisator symbol and structure

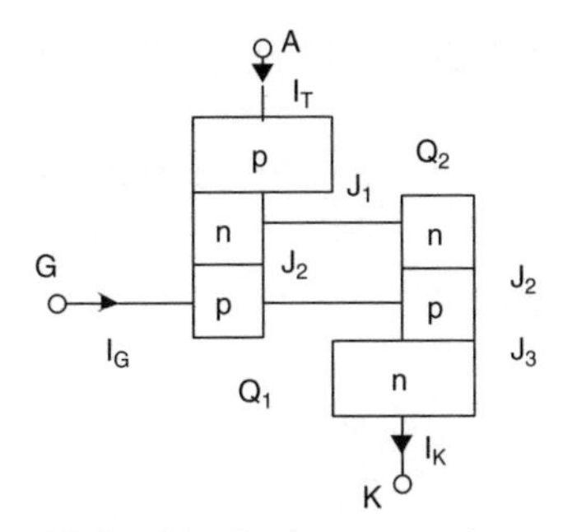

(b) Combination by two transistors

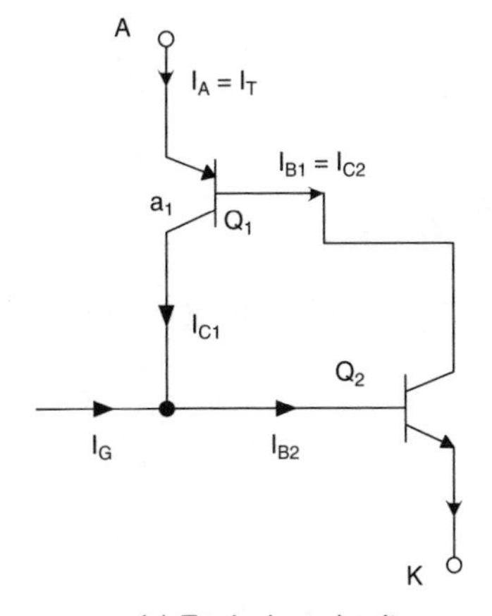

(c) Equivalent circuit

FIGURE 6.14
Thyristor.

6.6.1 Single-Phase Half-Wave Rectifier with Resistive Load

A single-phase half-wave thyristor rectifier with resistive load is shown in Figure 6.15. The firing angle α can be set in the range:

$$0 < \alpha < \pi$$

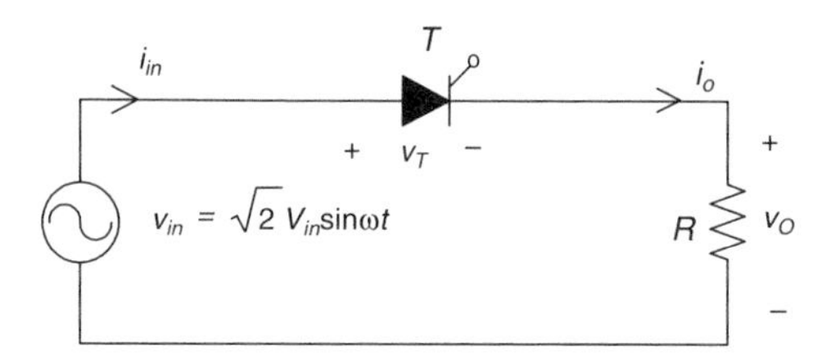

FIGURE 6.15
Single-phase half-wave thyristor rectifier with a resistor (R).

As before, the input voltage is a sinusoidal wave:

$$v_{in}(t) = \sqrt{2}V_{in}\sin\omega t \tag{6.1}$$

The output voltage and current are

$$v_O(t) = \begin{cases} \sqrt{2}V_{in}\sin\omega t & \alpha \le \omega t \le \pi \\ 0 & \pi < \omega t < 2\pi + \alpha \end{cases} \tag{6.51}$$

$$i_{in}(t) = i_O(t) = \begin{cases} \dfrac{\sqrt{2}V_{in}}{R}\sin\omega t & \alpha \le \omega t \le \pi \\ 0 & \pi < \omega t < 2\pi + \alpha \end{cases} \tag{6.52}$$

Their average values are

$$V_{O-av} = 0.225\,V_{in}(1+\cos\alpha) \tag{6.53}$$

$$I_{O-av} = 0.225\frac{V_{in}}{R}(1+\cos\alpha) \tag{6.54}$$

6.6.2 Single-Phase Half-Wave Thyristor Rectifier with Inductive Load

A single-phase half-wave thyristor rectifier with inductive load ($R + L$) is shown in Figure 6.16. The load time constant is $\tau = L/R$. Hence the load impedance Z is

$$Z = R + j\omega L = \sqrt{R^2 + (\omega L)^2}\ \angle\,\phi \tag{6.55}$$

where

$$\phi = \tan^{-1}\frac{\omega L}{R}$$

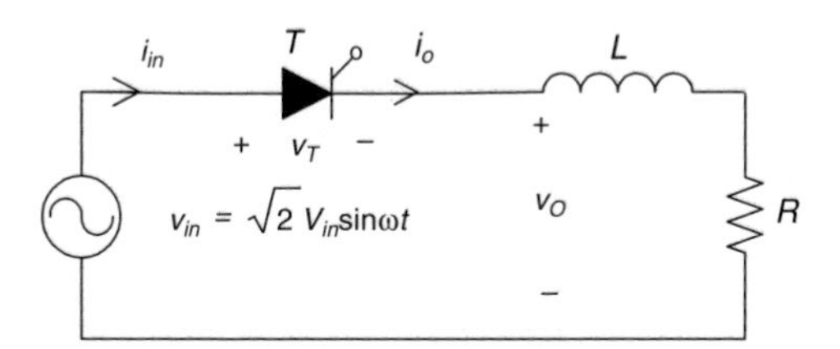

FIGURE 6.16
Single-phase half-wave thyristor rectifier with inductive load ($R + L$).

The output voltage and current are

$$v_O(t) = \begin{cases} \sqrt{2}V_{in}\sin\omega t & \alpha \le \omega t \le \beta \\ 0 & \beta < \omega t < 2\pi + \alpha \end{cases} \tag{6.56}$$

$$i_{in}(t) = i_O(t) = \begin{cases} \dfrac{\sqrt{2}V_{in}}{|Z|}\sin(\omega t - \phi) + \sin(\alpha - \phi)e^{-\frac{(\beta-\alpha)/\omega}{L/R}} & \alpha \le \omega t \le \beta \\ 0 & \beta < \omega t < 2\pi + \alpha \end{cases} \tag{6.57}$$

where β is the extinction angle and is determined by

$$\sin(\beta - \phi) = \sin(\alpha - \phi)e^{-\frac{\beta-\alpha}{\tan\phi}} \tag{6.58}$$

Their average values are

$$V_{O-av} = 0.225\,V_{in}(\cos\alpha - \cos\beta) \tag{6.59}$$

$$I_{O-av} = 0.225\frac{V_{in}}{R}(\cos\alpha - \cos\beta) \tag{6.60}$$

6.6.3 Single-Phase Half-Wave Thyristor Rectifier with Pure Inductive Load

A single-phase half-wave thyristor rectifier with pure inductive load (L) is shown in Figure 6.6. The load is an inductor L only. Therefore,

$$Z = j\omega L = \omega L \angle \phi \tag{6.61}$$

where

$$\phi = \frac{\pi}{2}$$

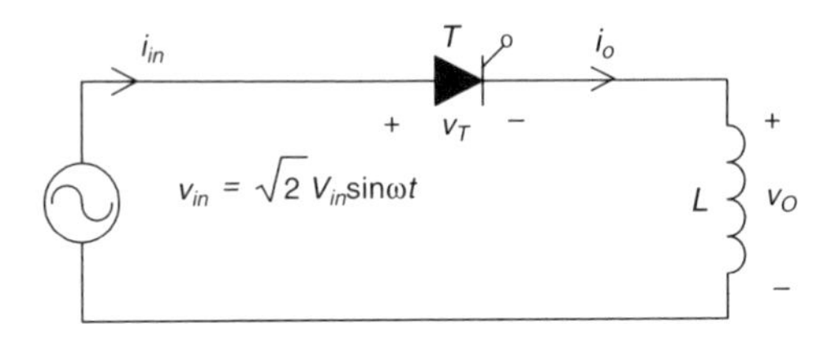

FIGURE 6.17
Single-phase half-wave thyristor rectifier with pure inductive load (L).

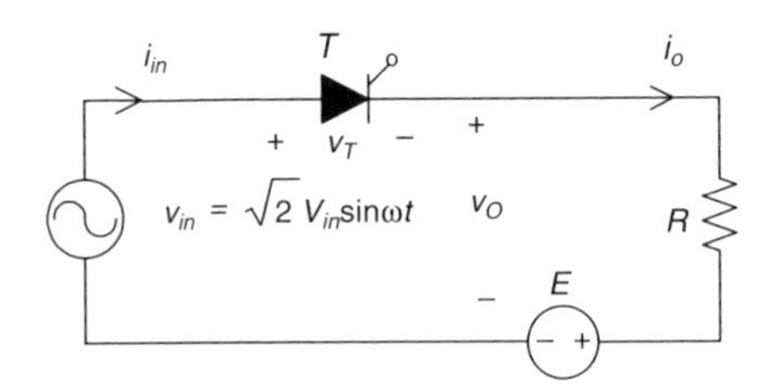

FIGURE 6.18
Single-phase half wave thyristor rectifier with EMF plus resistor (R + EMF).

The output voltage and current are

$$v_O(t) = \begin{cases} \sqrt{2}V_{in}\sin\omega t & \alpha \le \omega t \le 2\pi - \alpha \\ 0 & 2\pi - \alpha < \omega t < 2\pi + \alpha \end{cases} \tag{6.62}$$

$$i_{in}(t) = i_O(t) = \begin{cases} \dfrac{\sqrt{2}V_{in}}{|Z|}(con\alpha - con\omega t) & \alpha \le \omega t \le 2\pi - \alpha \\ 0 & 2\pi - \alpha < \omega t < 2\pi + \alpha \end{cases} \tag{6.63}$$

Their average values are

$$V_{O-av} = 0 \tag{6.64}$$

$$I_{O-av} = \frac{\sqrt{2}V_{in}}{\omega L\pi}[(\pi - \alpha) + \sin\alpha] \tag{6.65}$$

6.6.4 Single-Phase Half-Wave Rectifier with Back EMF Plus Resistive Load

A single-phase half wave thyristor rectifier with an EMF plus resistor load (a resistor R plus an EMF) is shown in Figure 6.18. The EMF value is E, which is smaller than the input peak voltage $\sqrt{2}V_{in}$. Suppose an auxiliary parameter m:

$$m = E/\sqrt{2}V_{in} < 1$$

and

$$\eta = \sin^{-1} m = \sin^{-1} \frac{E}{\sqrt{2}V_{in}}$$

The circuit will be analyzed for the relationship of the output to input. The AC supply voltage is sinusoidal.

Input voltage is $$v_{in}(t) = \sqrt{2}V_{in} \sin \omega t \tag{6.1}$$

The thyristor conduction condition requires firing angle $\alpha \geq \eta$. Hence, output voltage is

$$v_O(t) = \begin{cases} \sqrt{2}V_{in} \sin \omega t & \alpha \leq \omega t \leq (\pi - \eta) \\ E & (\pi - \eta) < \omega t < (2\pi + \alpha) \end{cases} \tag{6.66}$$

The input wave is a sinusoidal waveform, the corresponding output is a partial sinusoidal waveform less than half-cycle. The output DC average voltage is

$$V_{O-av} = \frac{\sqrt{2}V_{in}}{2\pi}(\cos \eta + \cos \alpha) + \frac{E}{2\pi}(\alpha + \pi + \eta) > E \tag{6.67}$$

The input and output current waveform is no longer a sinusoidal waveform.

$$i_{in}(t) = i_O(t) = \begin{cases} \frac{1}{R}(\sqrt{2}V_{in} \sin \omega t - E) & \alpha \leq \omega t \leq (\pi - \eta) \\ 0 & (\pi - \eta) < \omega t < (2\pi + \alpha) \end{cases} \tag{6.68}$$

The input and output current average value is

$$I_{O-av} = \frac{\sqrt{2}V_{in}}{2\pi R}[\cos \eta + \cos \alpha - m(\pi - \eta - \alpha)] \tag{6.69}$$

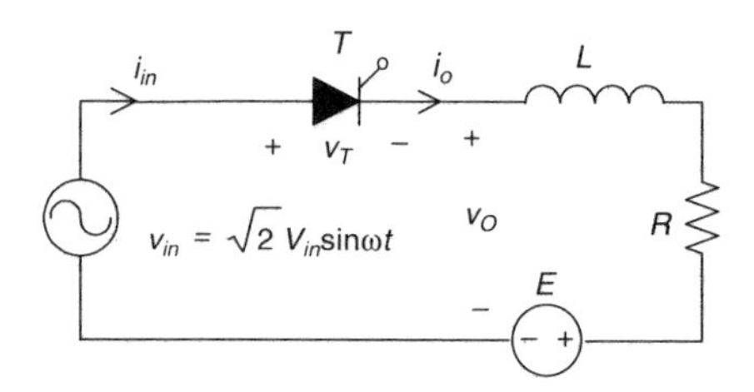

FIGURE 6.19
Single-phase half wave thyristor rectifier with EMF plus resistor and inductor ($EMF + R + L$).

6.6.5 Single-Phase Half-Wave Rectifier with Back EMF Plus Inductive Load

A single-phase half wave thyristor rectifier with an EMF plus inductive load (an EMF plus a resistor R and an inductor L) is shown in Figure 6.19. The EMF value is E, which is smaller than the input peak voltage $\sqrt{2}V_{in}$. Suppose an auxiliary parameter m:

$$m = E/\sqrt{2}V_{in} < 1$$

and

$$\eta = \sin^{-1} m = \sin^{-1}\frac{E}{\sqrt{2}V_{in}}$$

As before, the load Z is

$$Z = R + j\omega L = \sqrt{R^2 + (\omega L)^2}\ \angle\ \phi$$

where

$$|Z| = \sqrt{R^2 + (\omega L)^2} \quad \text{and} \quad \phi = \tan^{-1}\frac{\omega L}{R}$$

the load time constant $\tau = L/R$.

The circuit will be analyzed for the relationship of the output to input. The AC supply voltage is sinusoidal.

Input voltage is
$$v_{in}(t) = \sqrt{2}V_{in}\sin\omega t \tag{6.1}$$

Since the inductor causes the current continuity, the thyristor conduction angular length is γ, in which usually $\alpha + \gamma \geq \pi$. The thyristor conduction condition requires firing angle $\alpha \geq \eta$. Hence, output voltage is

$$v_O(t) = \begin{cases} \sqrt{2}V_{in}\sin\omega t & \alpha \le \omega t \le (\alpha+\gamma) \\ E & (\alpha+\gamma) < \omega t < (2\pi+\alpha) \end{cases} \tag{6.70}$$

Where γ is determined by:

$$e^{-\frac{\gamma}{\omega L/R}} = \frac{\frac{m}{\cos\phi} - \sin(\alpha+\gamma-\phi)}{\frac{m}{\cos\phi} - \sin(\alpha-\phi)} \tag{6.71}$$

The input wave is sinusoidal waveform, the corresponding output is a partial sinusoidal waveform less than half-cycle. The output DC average voltage is

$$V_{O-av} = E \tag{6.72}$$

The input and output current waveform is no longer a sinusoidal waveform.

$$i_{in}(t) = i_O(t)$$

$$= \begin{cases} \left(\begin{array}{l} \frac{\sqrt{2}V_{in}}{|Z|}[\sin(\omega t-\phi) - \{\frac{m}{con\phi} \\ -[\frac{m}{con\phi} - \sin(\alpha-\phi)]e^{-\frac{\gamma}{\omega L/R}}\}] \end{array} \right) & \alpha \le \omega t \le (\alpha+\gamma) \\ 0 & (\alpha+\gamma) < \omega t < (2\pi+\alpha) \end{cases} \tag{6.73}$$

The input and output current average value is

$$I_{O-av} = \frac{\sqrt{2}V_{in}}{2\pi R}[\cos\alpha - \cos(\alpha+\gamma) - m\gamma] \tag{6.74}$$

6.6.6 Single-Phase Half-Wave Rectifier with Back EMF Plus Pure Inductor

A single-phase half wave thyristor rectifier with an EMF plus a pure inductor L is shown in Figure 6.20. The EMF value is E, which is smaller than the input peak voltage $\sqrt{2}V_{in}$. Suppose an auxiliary parameter m:

$$m = E/\sqrt{2}V_{in} < 1$$

and

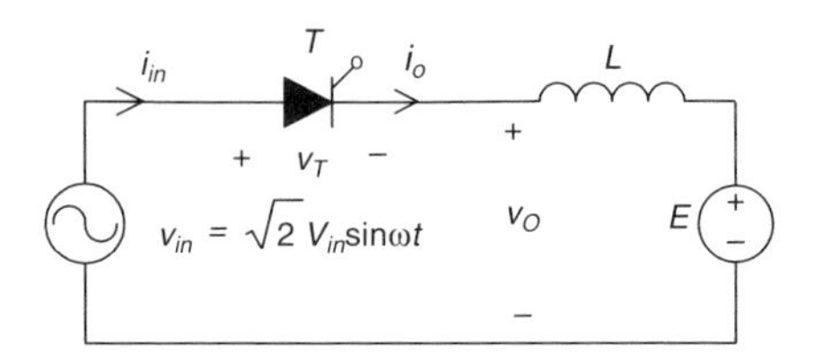

FIGURE 6.20
Single-phase half wave thyristor rectifier with EMF plus inductor (*EMF* + *L*).

$$\eta = \sin^{-1} m = \sin^{-1}\frac{E}{\sqrt{2}V_{in}}$$

The load Z is

$$Z = j\omega L = \omega L \angle \phi$$

where

$$\phi = \frac{\pi}{2}$$

The circuit will be analyzed for the relationship of the output to input. The AC supply voltage is sinusoidal.

Input voltage is
$$v_{in}(t) = \sqrt{2}V_{in}\sin\omega t \tag{6.1}$$

Since the inductor causes the current continuity, the thyristor conduction angular length is γ, which is usually $\alpha + \gamma \geq \pi$. The thyristor conduction condition requires firing angle $\alpha \geq \eta$. Hence, output voltage is

$$v_O(t) = \begin{cases} \sqrt{2}V_{in}\sin\omega t & \alpha \leq \omega t \leq (\alpha+\gamma) \\ E & (\alpha+\gamma) < \omega t < (2\pi+\alpha) \end{cases} \tag{6.75}$$

where γ is determined by:

$$\gamma = \frac{con\alpha - \cos(\alpha+\gamma)}{m} \tag{6.76}$$

The input wave is a sinusoidal waveform, the corresponding output is a partial sinusoidal waveform less than half-cycle. The output DC average voltage is

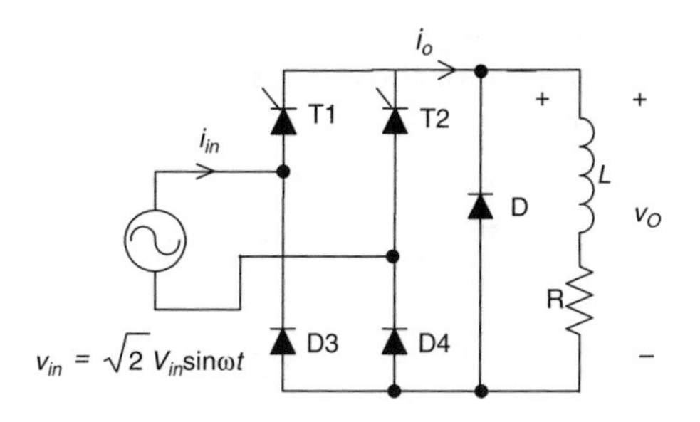

FIGURE 6.21
Single-phase full-wave semicontrolled thyristor rectifier with inductive load ($L + R$) plus a free-wheeling diode D.

$$V_{O-av} = E \tag{6.77}$$

The input and output current waveform is no longer a sinusoidal waveform.

$$i_{in}(t) = i_O(t)$$

$$= \begin{cases} \dfrac{\sqrt{2}V_{in}}{\omega L}[\cos\alpha - \cos\omega t - m(\omega t - \alpha)] & \alpha \le \omega t \le (\alpha + \gamma) \\ 0 & (\alpha + \gamma) < \omega t < (2\pi + \alpha) \end{cases} \tag{6.78}$$

The input and output current average value is

$$I_{O-av} = \frac{\sqrt{2}V_{in}}{2\pi\omega L}[\gamma\cos\alpha + \sin\alpha - \sin(\alpha+\gamma) - \frac{m\gamma^2}{2} + m\alpha\gamma] \tag{6.79}$$

6.6.7 Single-Phase Full-Wave Semicontrolled Rectifier with Inductive Load

A single-phase full-wave semicontrolled rectifier with inductive load (an inductor L and a resister R) plus a free-wheeling diode D, is shown in Figure 6.21. This rectifier is operating in quadrant 1 only because of the free-wheeling diode D. The rectified output waveform is repeating in the frequency 2ω. The load Z is

$$Z = R + j\omega L = \sqrt{R^2 + (\omega L)^2}\ \angle\ \phi$$

where

$$|Z| = \sqrt{R^2 + (\omega L)^2} \quad \text{and} \quad \phi = \tan^{-1}\frac{\omega L}{R}$$

the load time constant $\tau = L/R$.

The circuit will be analyzed for the relationship of the output to input. The AC supply voltage is sinusoidal.

Input voltage is $$v_{in}(t) = \sqrt{2}V_{in} \sin \omega t \tag{6.1}$$

Output voltage is $$v_O(t) = \begin{cases} \sqrt{2}V_{in} \sin \omega t & \alpha \le \omega t \le \pi \\ 0 & \pi < \omega t < (\pi + \alpha) \end{cases} \tag{6.80}$$

The output DC average voltage is

$$V_{O-av} = \frac{\sqrt{2}V_{in}}{\pi}(1 + \cos \alpha) \tag{6.81}$$

If the inductance is large enough and load time constant $\tau = L/R$ is larger than the half-cycle $T/2 = 1/2f$, and the input and output current waveform can be considered constant.

$$i_O(t) = \begin{cases} \dfrac{\sqrt{2}V_{in}}{(\pi - \alpha)R}(1 + \cos \alpha) & \alpha \le \omega t \le \pi \\ 0 & \pi < \omega t < (\pi + \alpha) \end{cases} \tag{6.82}$$

The input and output current average value is

$$I_{O-av} = \frac{\sqrt{2}V_{in}}{\pi R}(1 + \cos \alpha) \tag{6.83}$$

6.6.8 Single-Phase Full-Controlled Rectifier with Inductive Load

A single-phase full-wave semicontrolled thyristor rectifier with inductive load (an inductor L and a resister R) is shown in Figure 6.22. This rectifier is operating in quadrants I and IV. The rectified output waveform is repeating in the frequency 2ω. If the inductance is large enough, the output current can be constant. The load Z is

$$Z = R + j\omega L = \sqrt{R^2 + (\omega L)^2} \angle \phi$$

where

$$|Z| = \sqrt{R^2 + (\omega L)^2} \quad \text{and} \quad \phi = \tan^{-1}\frac{\omega L}{R}$$

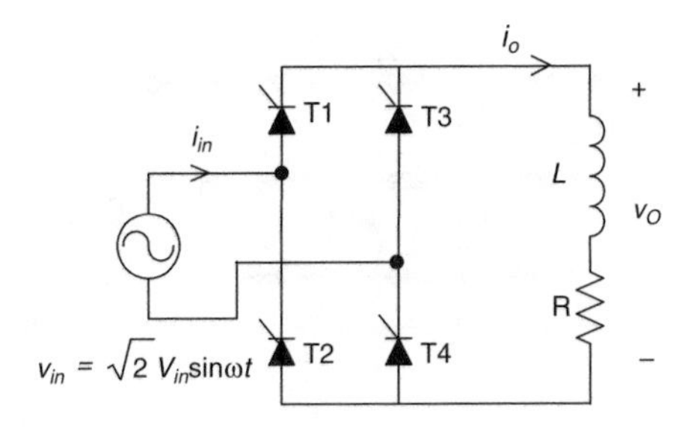

FIGURE 6.22
Single-phase full-wave semicontrolled thyristor rectifier with inductive load ($L + R$).

the load time constant $\tau = L/R >> T/2 = 1/2f$.

The circuit will be analyzed for the relationship of the output to input. The AC supply voltage is sinusoidal.

Input voltage is $$v_{in}(t) = \sqrt{2}V_{in}\sin\omega t \tag{6.1}$$

Output voltage is $$v_O(t) = \sqrt{2}V_{in}\sin\omega t \qquad \alpha \le \omega t \le (\pi + \alpha) \tag{6.84}$$

The output DC average voltage is

$$V_{O-av} = \frac{2\sqrt{2}V_{in}}{\pi}\cos\alpha \tag{6.85}$$

If the inductance is large enough and load time constant $\tau = L/R$ is larger than the half-cycle $T/2 = 1/2f$, the input and output current waveform can be considered constant.

$$i_O(t) = \frac{2\sqrt{2}V_{in}}{\pi R}\cos\alpha \tag{6.86}$$

The input and output current average value is

$$I_{O-av} = \frac{\sqrt{2}V_{in}}{\pi R}(1+\cos\alpha) \tag{6.87}$$

6.6.9 Three-Phase Half-Wave Rectifier with Resistive Load

A three-phase half-wave thyristor rectifier with a resister R is shown in Figure 6.23. This rectifier is operating in quadrant I only. The rectified output waveform is repeating in the frequency 3ω. The circuit will be

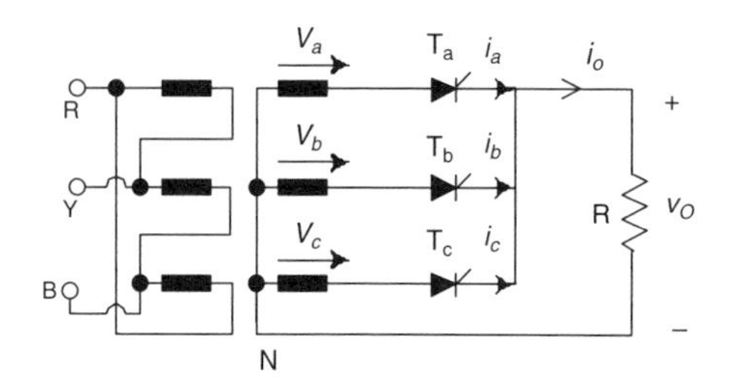

FIGURE 6.23
Three-phase half-wave semicontrolled thyristor rectifier with resister.

analyzed for the relationship of the output to input. The AC supply voltage is sinusoidal.

Input line-to-neutral (between phase *a* to *N*) voltage is

$$v_{aN}(t) = \sqrt{2}V_{in}\sin(\omega t - \frac{\pi}{6}) \tag{6.88}$$

Output voltage is
$$v_O(t) = \begin{cases} \sqrt{2}V_{in}\sin\omega t & \alpha \le \omega t \le \dfrac{2\pi}{3} \\ 0 & \omega t > \dfrac{2\pi}{3} \end{cases} \tag{6.89}$$

The firing angle α starts from the phase cross point ωt = 30°. Each thyristor's maximum conduction period is 120°. Possible firing angle range is

$$0 \le \alpha \le 150°$$

The output DC average voltage is

$$V_{O-av} = \begin{cases} \dfrac{3\sqrt{6}V_{in}}{2\pi}\cos\alpha & \alpha \le \dfrac{\pi}{6} \\ \sqrt{6}V_{in}\sqrt{\dfrac{1}{6} + \dfrac{\sqrt{3}}{8\pi}\cos 2\alpha} & \alpha > \dfrac{\pi}{6} \end{cases} \tag{6.90}$$

The output current waveform is a partial sinusoidal wave.

$$v_O(t) = \begin{cases} \dfrac{\sqrt{2}V_{in}}{R}\sin\omega t & \alpha \le \omega t \le \dfrac{2\pi}{3} \\ 0 & \omega t > \dfrac{2\pi}{3} \end{cases} \tag{6.91}$$

The output current average value is

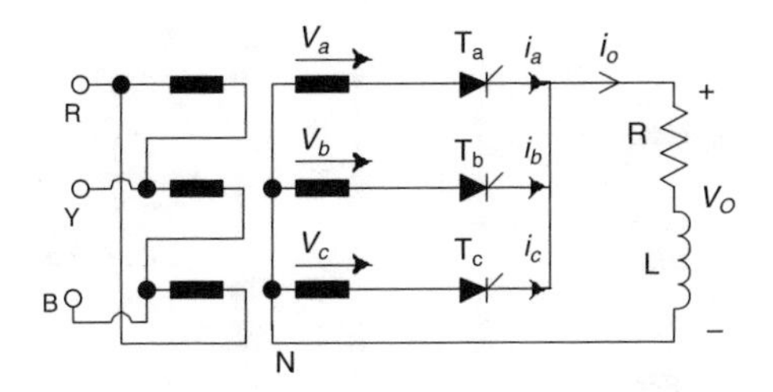

FIGURE 6.24
Three-phase half-wave semicontrolled thyristor rectifier with inductive load ($L + R$).

$$V_{O-av} = \begin{cases} \dfrac{3\sqrt{6}V_{in}}{2\pi R}\cos\alpha & \alpha \le \dfrac{\pi}{6} \\ \dfrac{\sqrt{6}V_{in}}{R}\sqrt{\dfrac{1}{6}+\dfrac{\sqrt{3}}{8\pi}\cos 2\alpha} & \alpha > \dfrac{\pi}{6} \end{cases} \tag{6.92}$$

6.6.10 Three-Phase Half-Wave Thyristor Rectifier with Inductive Load

A three-phase half-wave thyristor rectifier with inductive load (an inductor L and a resister R) is shown in Figure 6.24. This rectifier is operating in quadrants I and IV. The rectified output waveform is repeating in the frequency 3ω. If the inductance is large enough, the output current can be constant. The load Z is

$$Z = R + j\omega L = \sqrt{R^2 + (\omega L)^2}\ \angle\ \phi$$

where

$$|Z| = \sqrt{R^2 + (\omega L)^2} \quad \text{and} \quad \phi = \tan^{-1}\frac{\omega L}{R}$$

the load time constant $\tau = L/R >> T/3 = 1/3f$.

The circuit will be analyzed for the relationship of the output to input. The AC supply voltage is sinusoidal.

Input line-to-neutral (between phase a to N) voltage is

$$v_{aN}(t) = \sqrt{2}V_{in}\sin(\omega t - \frac{\pi}{6}) \tag{6.88}$$

Output voltage is $v_O(t) = \sqrt{2}V_{in}\sin\omega t \quad \alpha \le \omega t \le (\frac{2\pi}{3} + \alpha)$ (6.93)

The firing angle α starts from the phase cross point $\omega t = 30°$. Possible firing angle range is

$$0 \le \alpha \le 180°$$

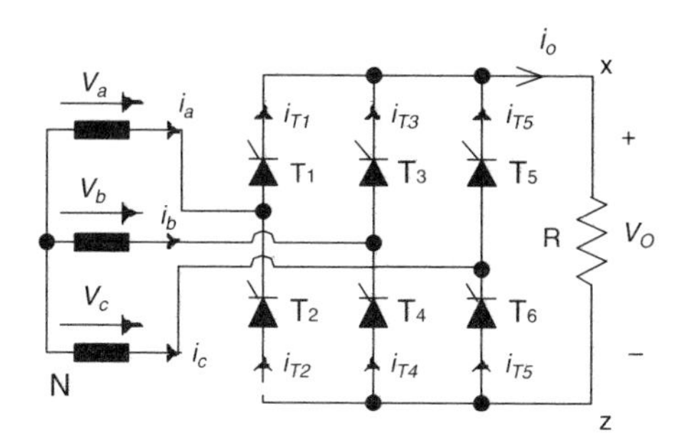

FIGURE 6.25
Three-phase full-wave thyristor rectifier with resister R.

The output DC average voltage is

$$V_{O-av} = \frac{3\sqrt{6}V_{in}}{2\pi}\cos\alpha \tag{6.94}$$

If the inductance is large enough and load time constant $\tau = L/R$ is larger than the half-cycle $T/2 = 1/2f$, the output current waveform can be considered constant.

$$i_O(t) = \frac{3\sqrt{6}V_{in}}{2\pi R}\cos\alpha \tag{6.95}$$

The output current average value is

$$I_{O-av} = \frac{3\sqrt{6}V_{in}}{2\pi R}\cos\alpha \tag{6.96}$$

6.6.11 Three-Phase Full-Wave Thyristor Rectifier with Resistive Load

A three-phase full-wave thyristor rectifier with a resister R is shown in Figure 6.25. This rectifier is operating in quadrant I only. The rectified output waveform is repeating in the frequency 6ω. The circuit will be analyzed for the relationship of the output to input. The AC supply voltage is sinusoidal.

Input line-to-line (between phase a to c) voltage is

$$v_{ac}(t) = \sqrt{6}V_{in}\sin(\omega t - \frac{\pi}{6}) \tag{6.88}$$

Output voltage is $$v_O(t) = \begin{cases} \sqrt{6}V_{in}\sin\omega t & \alpha \le \omega t \le \frac{2\pi}{3} \\ 0 & \omega t > \frac{2\pi}{3} \end{cases} \tag{6.97}$$

The firing angle α starts from the phase cross point ωt = 30°. Each thyristor maximum conduction period is 120°. Possible firing angle range is

$$0 \le \alpha \le 150°$$

The output DC average voltage is

$$V_{O-av} = \begin{cases} \dfrac{3\sqrt{6}V_{in}}{\pi}\cos\alpha & \alpha \le \dfrac{\pi}{6} \\ 2\sqrt{6}V_{in}\sqrt{\dfrac{1}{6} + \dfrac{\sqrt{3}}{8\pi}\cos 2\alpha} & \alpha > \dfrac{\pi}{6} \end{cases} \tag{6.98}$$

The output current waveform is a partial sinusoidal wave.

$$v_O(t) = \begin{cases} \dfrac{\sqrt{6}V_{in}}{R}\sin\omega t & \alpha \le \omega t \le \dfrac{2\pi}{3} \\ 0 & \omega t > \dfrac{2\pi}{3} \end{cases} \tag{6.99}$$

The output current average value is

$$I_{O-av} = \begin{cases} \dfrac{3\sqrt{6}V_{in}}{2\pi R}\cos\alpha & \alpha \le \dfrac{\pi}{6} \\ \dfrac{\sqrt{6}V_{in}}{R}\sqrt{\dfrac{1}{6} + \dfrac{\sqrt{3}}{8\pi}\cos 2\alpha} & \alpha > \dfrac{\pi}{6} \end{cases} \tag{6.100}$$

6.6.12 Three-Phase Full-Wave Thyristor Rectifier with Inductive Load

A three-phase full-wave thyristor rectifier with inductive load (an inductor L and a resister R) is shown in Figure 6.26. This rectifier is operating in quadrants I and II. The rectified output waveform is repeating in the frequency 6ω. If the inductance is large enough, the output current can be constant. The load Z is

$$Z = R + j\omega L = \sqrt{R^2 + (\omega L)^2} \angle \phi$$

where

$$|Z| = \sqrt{R^2 + (\omega L)^2} \quad \text{and} \quad \phi = \tan^{-1}\frac{\omega L}{R}$$

the load time constant $\tau = L/R >> T/3 = 1/3f$

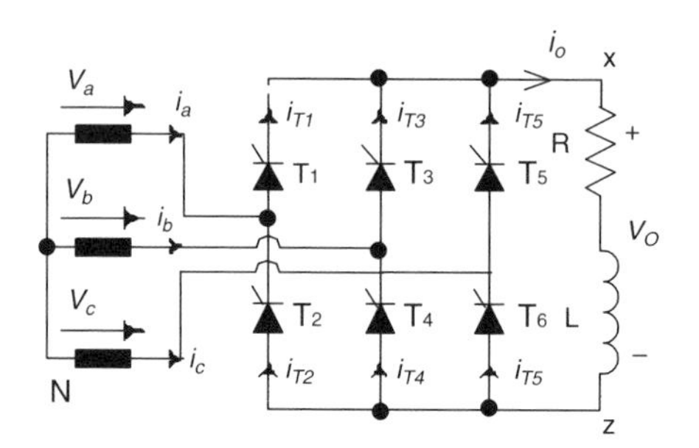

FIGURE 6.26
Three-phase full-wave thyristor rectifier with inductive load ($L + R$).

The circuit will be analyzed for the relationship of the output to input. The AC supply voltage is sinusoidal.

Input line-to-line (between phase a to c) voltage is

$$v_{ac}(t) = \sqrt{6}V_{in}\sin(\omega t - \frac{\pi}{6}) \tag{6.88}$$

Output voltage is $v_O(t) = \sqrt{6}V_{in}\sin\omega t \qquad \alpha \le \omega t \le (\frac{2\pi}{3} + \alpha)$ (6.101)

The firing angle α starts from the phase cross point ωt = 30°. Possible firing angle range is

$$0 \le \alpha \le 180°$$

The output DC average voltage is

$$V_{O-av} = \frac{3\sqrt{6}V_{in}}{\pi}\cos\alpha \tag{6.102}$$

If the inductance is large enough and load time constant $\tau = L/R$ is large than the half-cycle $T/6 = 1/6f$, the output current waveform can be considered constant.

$$i_O(t) = \frac{3\sqrt{6}V_{in}}{\pi R}\cos\alpha \tag{6.103}$$

The output current average value is

$$I_{O-av} = \frac{3\sqrt{6}V_{in}}{\pi R}\cos\alpha \tag{6.104}$$

Bibliography

Rectifier Applications Handbook, 3rd ed., Phoenix, Ariz: Motorola, Inc., 1993.

Bird, B.M. and King, K.G., *An Introduction to Power Electronics*, New York, Prentice-Hall, 1983.

Dwyer, R. and Mueller, D., Selection of Transformers for Commercial Building, in *Proceedings of IEEE-IAS'92*, U.S., 1992, p. 1335.

Lee, Y.S., *Computer-Aided Analysis and Design of Switch-Mode Power Supplies*, New York: Marcel Dekker, Inc., 1993.

Thorborg, K., *Power Electronics*, Prentice-Hall International Ltd., UK, 1988.

Todd, P., *UC3854 Controlled Power Factor Correction Circuit Design*, Application Note U-134, Unitrode Corp, 1995.

7

Control Circuit: EMI and Application Examples of DC/DC Converters

7.1 Introduction

During investigation of DC/DC prototypes and their characteristics, much attention is paid to the circuitry components of the converters. Actually, control components as auxiliary apparatus are important roles for DC/DC converter operation. For example, the PWM pulse train generator is used to yield the switching signal to all switches of DC/DC converters.

EMI, EMS, and EMC have to be considered during DC/DC converter design because they affect the converter and other equipment working operation heavily.

Some particular examples of DC/DC converters are presented in this chapter to demonstrate DC/DC converter application.

7.2 Luo-Resonator

The Luo-resonator is a pulse-width-modulated (PWM) signal generator, which produces the PWM pulse train switching signal used for DC/DC converters. This resonator consists of only three *operational amplifiers* (OA), and provides a pulse train of the switching signal to control static switch-on or switch-off with adjustable frequency f and conduction duty k. Luo-resonator can be re-integrated into an application specific integrated circuit (ASIC) to produce portable DC/DC converters.

The Luo-resonator is a high efficiency and simple circuit with easily adjusting frequency f and conduction duty k. Its circuit diagram is shown in Figure 7.1. It consists of three OAs named OA1 to 3 and auxiliary. These three 741-type OAs are integrated in a chip TL074 (which contains four OAs). Two potentiometers are applied to adjust the frequency f and conduction duty k.

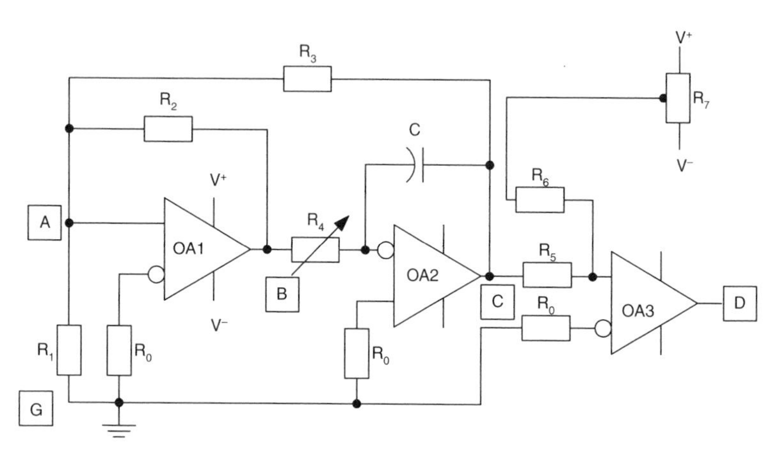

FIGURE 7.1
Luo-resonator.

The analysis of the Luo-resonator is performed under the assumption that the operational amplifier is ideal:

1. Its open-loop gain is infinity
2. Its input impedance is infinity and output impedance is zero
3. Its output voltage positive and negative maximum values are equal to the power supply voltages

7.2.1 Circuit Explanation

Type-741 OA can work with a ±3 to ±18 V power supplies, which are marked $V+$, G, and $V-$ with $|V-| = V+$. OA2 in Figure 7.1 acts as the integration operation, its output V_C is a triangle waveform with regulated frequency $f = 1/T$ controlled by potentiometer R_4. OA1 acts as a resonant operation, its output V_B is a square-waveform with the frequency f. OA3 acts as a comparator, its output V_D is a square-waveform pulse train with regulated conduction duty k controlled by R_7.

First, the output voltage of OA1 maintained as $V_B = V+$. In the meantime V_B inputs to OA2 via R_4. Because of the capacitor C, the output voltage V_C of OA2 decreases toward $V-$ with the slope $-1/R_4C$. Voltage V_C feeds back to OA1 negatively via R_3. Voltage V_A at point A changes from $(2mV+)/(1+m)$ downward to 0 in the period of $0 - 2mR_4C$. It then intends toward negative. It causes the OA1's output voltage $V_B = V-$ at $t = 2mR_4C$ and voltage V_A jumps to

$$\frac{2mV-}{1+m} \tag{7.1}$$

Therefore, the output voltage of OA1 jumps to $V_B = V-$. In the meantime V_B inputs to OA2 via R_4. Because of the capacitor C, the output voltage V_C of OA2 increases toward $V+$ with the slope $1/R_4C$. Voltage V_C feeds back to OA1 negatively via R_3. Voltage V_A at point A changes from $(2mV-)/(1+m)$ upwards to 0 in the period of $2mR_4C - 4mR_4C$. It then intends toward positive. It causes the OA1's output voltage $V_B = V+$ at $t = 4mR_4C$ and voltage V_A jumps to

$$\frac{2mV+}{1+m} \tag{7.2}$$

Voltage V_B takes the two values either $V+$ or $V-$.

Voltage V_C is a triangle waveform, and inputs to OA3. It compares with shift signal $V_{off\text{-}set}$ regulated by the potentiometer R_7 via R_6. When $V_{off\text{-}set} = 0$, OA3 yields its output voltage V_D as a pulse train with conduction duty $k = 0.5$. Positive $V_{off\text{-}set}$ shifts the zero-cross point of voltage V_C downward, hence, OA3 yields its output voltage V_D as a pulse train with conduction duty $k > 0.5$. Vice versa, negative $V_{off\text{-}set}$ shifts the zero-cross point of voltage V_C upward, hence, OA3 yields its output voltage V_D as a pulse train with conduction duty $k < 0.5$ as shown in Figure 7.2. Conduction duty k is controlled by $V_{off\text{-}set}$ via the potentiometer R_7.

7.2.2 Calculation Formulae

The calculation formulas are
Setting,

$$m = \frac{R_3}{R_2} \tag{7.3}$$

We obtain:

$$f = \frac{1}{T} = \frac{1}{4mR_4C} \tag{7.4}$$

and

$$k = 0.5 + \frac{R_5 V_{off-set}}{2R_6 V_+} \tag{7.5}$$

If the positive and negative maximum values of the shift voltage $V_{off\text{-}set}$ are $V+$ and $V-$, and $R_5 = R_6$, the value of the conduction duty k is in the range between 0 and 1.0. Considering the resistance tolerance is 5%, we usually choose that resistance R_6 is slightly smaller than resistance R_5.

This PWM pulse train V_D is applied to the DC/DC converter switch such as a transistor, MOSFET or IGBT via a coupling circuit. The voltage waveforms of $V_A - V_D$ are shown in Figure 7.2.

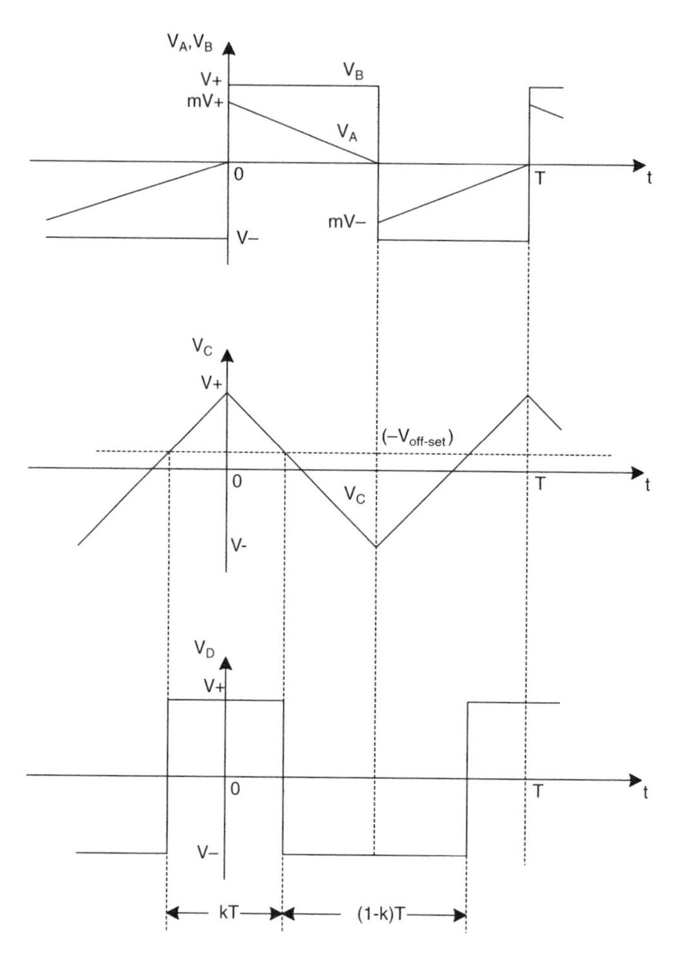

FIGURE 7.2
Voltage waveforms of Luo-resonator.

7.2.3 A Design Example

A Luo-resonator was designed as shown in Figure 7.2 with the component values:

$$R_0 = 10\ \text{k}\Omega;\ R_1 = R_2 = R_5 = 100\ \text{k}\Omega;\ R_3 = R_6 = 95\ \text{k}\Omega;$$
$$R_4 = 510\ \Omega \text{ to } 5.1\ \text{k}\Omega;\ R_7 = 20\ \text{k}\Omega \text{ and } C = 5.1\ \text{nF}$$

The results are $m = 0.95$, frequency $f = 10$ kHz to 100 kHz and conduction duty $k = 0$ to 1.0.

7.2.4 Discussion

Type 741 operational amplifier of chip TL074 has the frequency bandwidth of about 2 MHz. Its open-loop gain is only about 20 when this Luo-resonator works at $f = 100$ kHz. The waveform of V_C may be deformed slightly. It increases or decreases as an exponential curve, but linearly. However, the frequency f and conduction duty k of the output PWM pulse train are still adjustable.

Although real maximum positive and negative output voltages of all OA are slightly smaller than the power supply voltages $V+$ and $V-$, experimental results verified that Luo-resonator still works well. When power supply voltages change from ± 5 V to ± 18 V the variations of the frequency f and conduction duty k are less than 2%.

7.3 EMI, EMS, and EMC

Electromagnetic interference (EMI) generally exists in all electrical and electronic equipment, especially in all DC/DC converters. Since the switching frequency applied in DC/DC converter is high it causes significant EMI if carelessly designed. For the sake of providing the power quality, two objectives must be achieved. First, limit the radio frequency (RF) emissions that can be imposed on the power mains. Second, elaborate the current electromagnetic susceptibility (EMS) test methods with the goal of reducing the EMS of devices on the consumer end of the power grid.

Electromagnetic compatibility (EMC) has come a long way from the "black magic" approach in the early 1960s to an almost exact science today with its analytical methods, measurement techniques, and simulation software. Four decades ago, all existing handbooks on EMC could be counted on the fingers of one hand, but today they could occupy several shelves in a respectable library. This fact is caused by the significant reason that the applied frequency in 9 kHz to 30 MHz is much higher than those (1 kHz to 10 kHz) of 40 years ago. EMI is a serious problem in power electronic circuits because of their fast switching characteristics. Many countries have imposed EMC regulations that must be met before electronic products can be sold legally. Because of this fact, the vital importance of this problem in all equipment including DC/DC converters is recognized.

7.3.1 EMI/EMC Analysis

The recent investigation of international regulations on EMC has prompted active research in the study of EMI emission from switched-mode power converters, which are now indispensable components in modern electronic equipment such as computers. EMI study can be focused on three major elements, namely,

1. The EMI source (EMI emission)
2. The coupling path (EMI transmission)
3. The victim (EMI effect)

It is important to minimize the coupling path and to improve the EMI immunity. An effective solution for EMI suppression is to attack the problem at the EMI source. By studying power electronics circuits, the high dv/dt and di/dt involved in the switching operation of traditional hard-switched power electronics devices are the major source of **EMI emission**. Effectively reducing dv/dt and di/dt in DC/DC converters will largely attenuate the EMI emission in radio frequency (RF) radiation.

In order to improve the energy efficiency and reliability of power converters soft-switching techniques created in 1980s have been proposed to reduce

- The switching power losses across the power devices,
- The switching stress of switched-mode power electronics circuits.

Essentially, soft-switching techniques create a zero-voltage-switching (ZVS) and/or zero-current-switching (ZCS) conversion process for the power switched to turn-on and turn-off. Therefore, the instantaneous power losses across the main switches can be reduced or eliminated. Results published recently have confirmed the feasibility of such soft-switched operation.

Coupling path is a complex problem. EMI emission is created by the EMI source and radiated out by certain manner, then EMI reaches the receiver or victim. The EMI transferring process from the source to victim is called coupling path. It relies on the following factors

- The physical structure in both EMI source and victim
- The location and direction between EMI source and victim
- The transmission media (shielded or unshielded) between EMI source and victim
- The difference of the frequency bandwidths between EMI source and victim

Successfully cut, the coupling path can effectively reduce the EMI to other equipment (victim). Unfortunately, no matter how carefully it is done, EMI still affects all victims.

Victim is the equipment to be harmed by the EMI. In order to reduce the interference some effective measures can be taken

- Neat physical structure
- Distance from source
- Shielding equipment
- Large different frequency band from EMI source

We concentrate to reduce the EMI emission that may be created by DC/DC converters in further sub-sections.

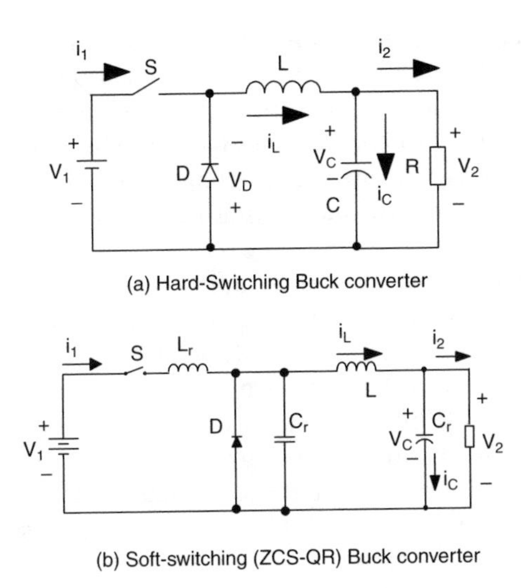

FIGURE 7.3
Hard-switching and soft-switching buck converter.

7.3.2 Comparison with Hard-Switching and Soft-Switching

Hard-switching converters usually have large rates of dv/dt and di/dt. For example, a hard-switching buck converter is shown in Figure 1.23a can be redrawn in Figure 7.3a and its corresponding soft-switching buck converter is shown in Figure 7.3b. The soft-switching buck converter is similar to its hard-switching circuit, except that it consists of the extra resonant components L_r and C_r. Since the values of the extra resonant components L_r and C_r are usually small, they can be carefully designed in the power devices (switch S and diode D) snubbers. Although a snubber circuit can reduce the EMI, it generally causes additional energy loss. Thus, resonant converters not only produce less EMI, but also exhibit lower energy loss than hard-switching converters with snubber circuit.

The resonant inductor L_r limits the initial current of the main switch S to provide zero-current condition during switch-on. The resonant process can provide another zero-current condition for switch-off operation. The resonant capacitor C_r can discharge through the antiparallel diode of the main switch S (it usually exists in the device), thus clamping the voltage across S to about 1 V for near-zero voltage turn-off of the main switch S.

7.3.3 Measuring Method and Results

An EMC analyzer is usually used to measure the EMI emission conducted and radiated. For example, an EMC analyzer HP 8591EM is widely applied in experiments to measure both conducted and radiated emissions. During the tests, the detector function was set to the Comite International Special des Perturbations Radioelectriques (CISPR) quasi-peak function. With the converter off, the frequency range of interest is swept to survey ambient environmental level. With the converter on, the signals measured are the ambient and the converter emission signals. The actual emission from the converter can be obtained by subtracting the ambient signal from the measured signals.

The basic converter components (L and C) in each set of hard-switched and soft-switched converters are identical and the two converters are tested under same load conditions. The magnitude of the inductor current and load current in the two converters of each type are essentially identical. The converters have no EMI filters and are not shielded. In addition, none of the converters has any enclosure. The background EMI was measured just before turning on each converter. Both the conducted EMI (from 50 kHz) and radiated EMI (from 50 kHz to 5 MHz) of the converters were recorded. In the radiated EMI measurement, the results have been corrected with the antenna. All converters were tested with an output power of about 55 W.

For both of the hard-switched and soft-switched buck converters, $V_{in} = 55$ V, $V_o = 20$ V, switching frequency $f = 50$ kHz and duty cycle $k = 0.4$, $L = 2.5$ mH, $C = 20$ μF and, $R = 7.5$ Ω. For the soft-switched buck converter, $L_r = 4$ μH and $C_r = 1$ μF. Figure 7.4 shows the voltage and current waveforms of the main switch S in the hard-switched buck converter. The corresponding current FFT spectrum is shown in Figure 7.5. The switching trajectory in this case is illustrated in Figure 7.6. Figure 7.9 shows the voltage and current waveforms of the main switch S in the soft-switched buck converter. The corresponding current FFT spectrum is shown in Figure 7.8. The switching trajectory in this case is illustrated in Figure 7.9. The L-shape of the trajectory confirms the main switch S in soft-switching state.

Comparison of Figure 7.4 and Figure 7.7 shows that the soft-switched voltage and current waveforms have very little transient ringing. The reverse recovery current of the diode in the soft-switched converter is also much less than in the hard-switched converter. The significant reduction of the transient ringing in the soft-switched converter results in much reduced dv/dt and di/dt.

Comparison of the FFT spectrums in Figure 7.5 and Figure 7.8 shows that the soft-switched current has much lower total harmonic distortion (THD) than hard-switched current. The FFT spectrum in Figure 7.8 shows the frequency bands are lower that that in Figure 7.5. Otherwise, amplitudes of all harmonics in Figure 7.8 are lower than those in Figure 7.5. From this fact, EMI emission is much lower than that of the corresponding soft-switching converter.

Comparison of the FFT spectrums in Figure 7.6 and Figure 7.9 shows that the switching trajectory shows that the conducted and radiated EMI emission from the hard-switched buck converter (together with the background noise) is larger than that from the soft-switched buck converter. It can be seen that

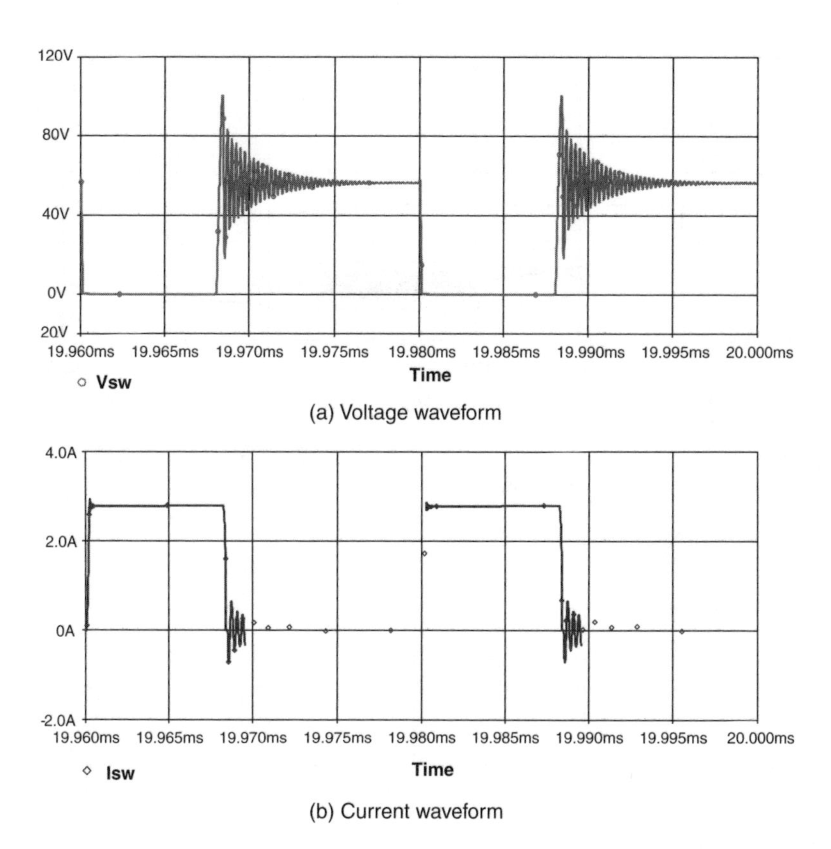

FIGURE 7.4
Voltage and current waveforms of hard-switching buck converter.

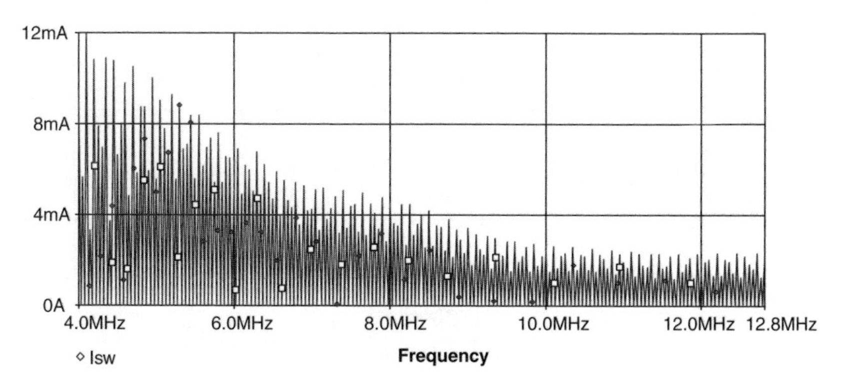

FIGURE 7.5
Current FFT spectrum of hard-switching buck converter.

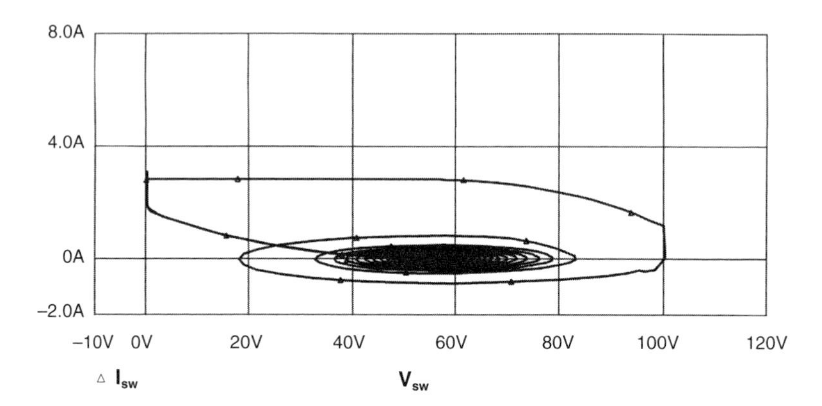

FIGURE 7.6
Switching trajectory of hard-switching buck converter.

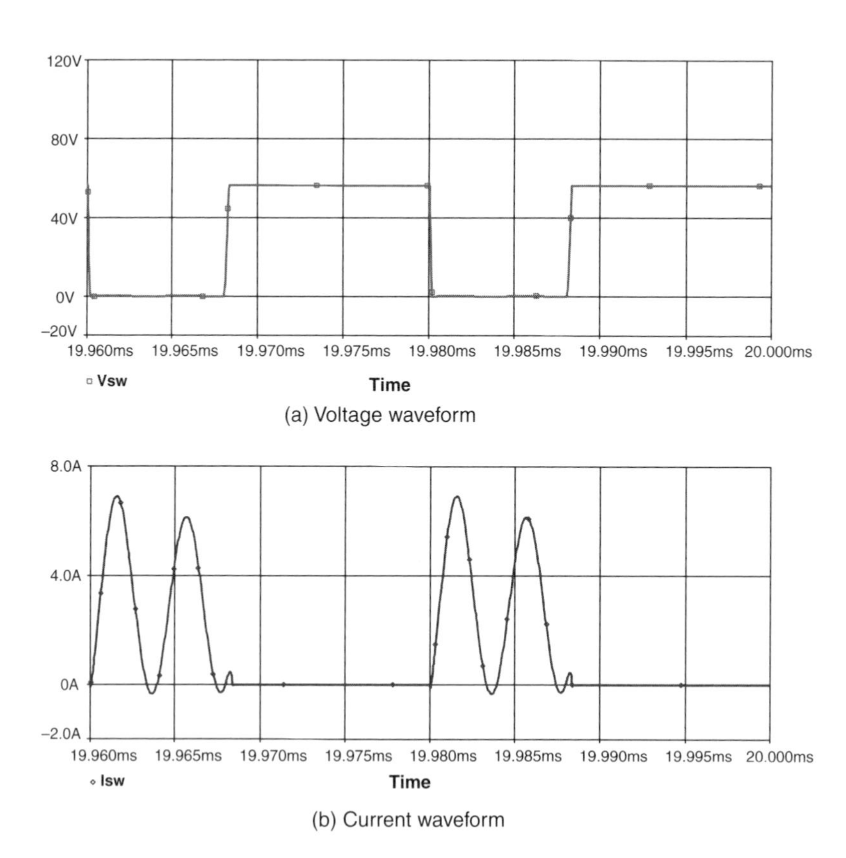

FIGURE 7.7
Voltage and current waveforms of soft-switching buck converter.

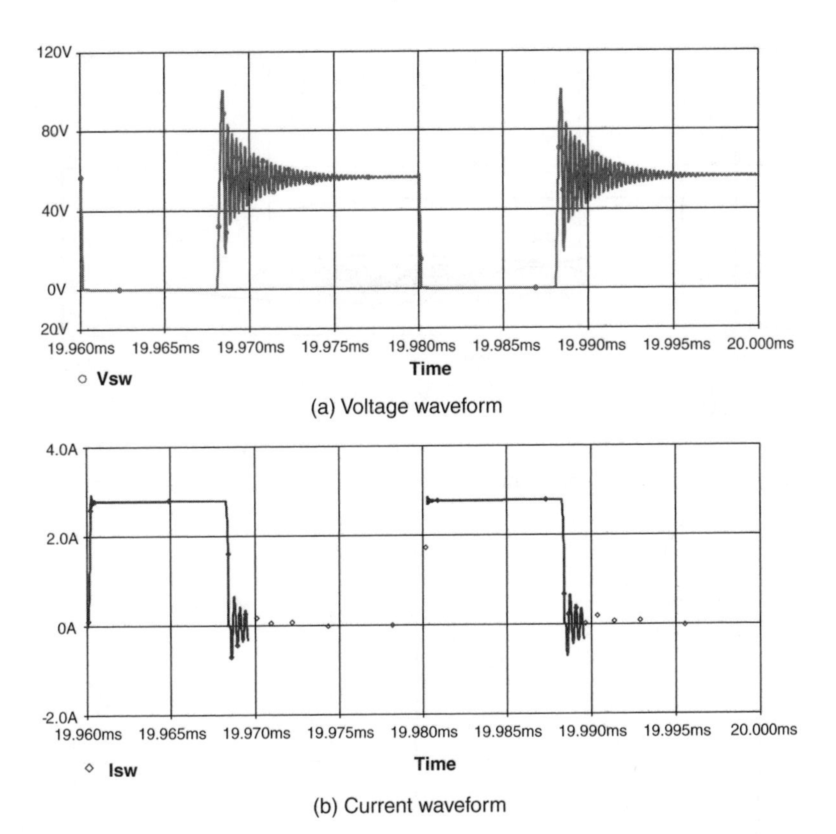

FIGURE 7.4
Voltage and current waveforms of hard-switching buck converter.

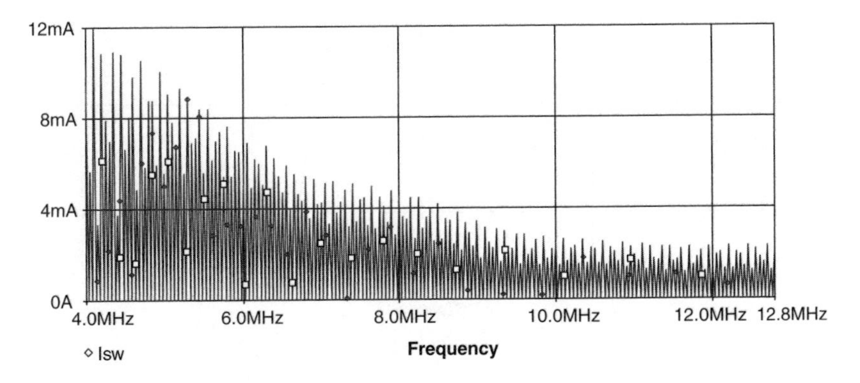

FIGURE 7.5
Current FFT spectrum of hard-switching buck converter.

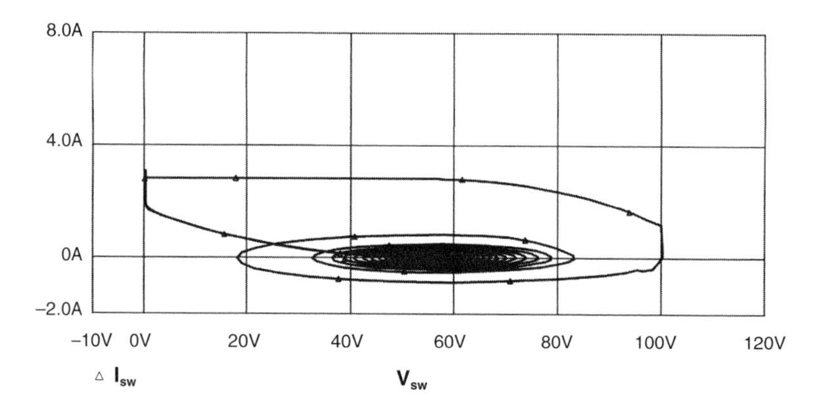

FIGURE 7.6
Switching trajectory of hard-switching buck converter.

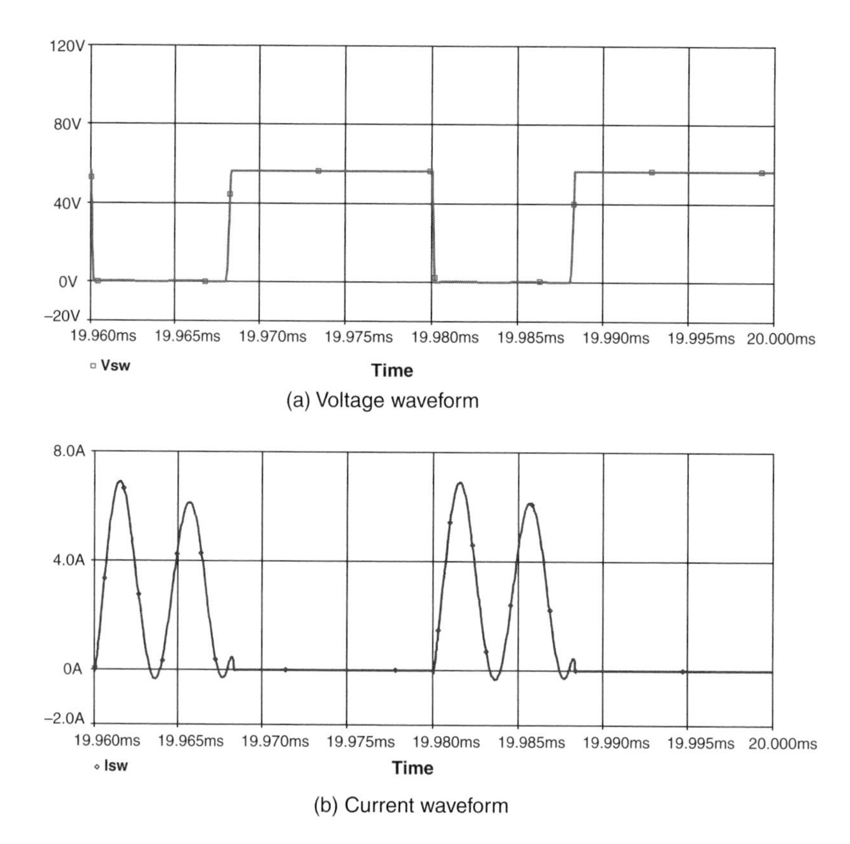

FIGURE 7.7
Voltage and current waveforms of soft-switching buck converter.

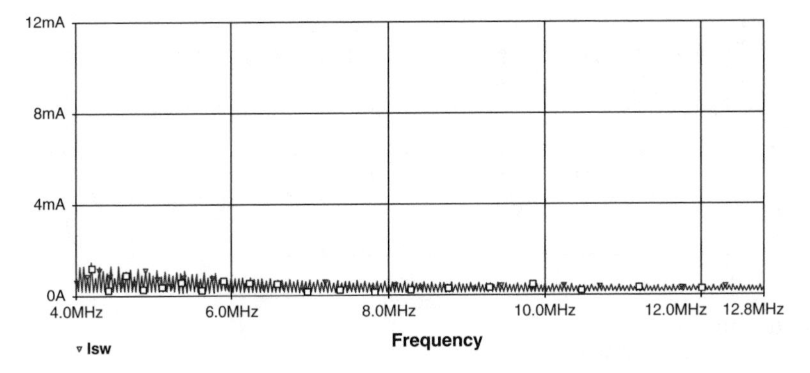

FIGURE 7.8
Current FFT spectrum of soft-switching buck converter.

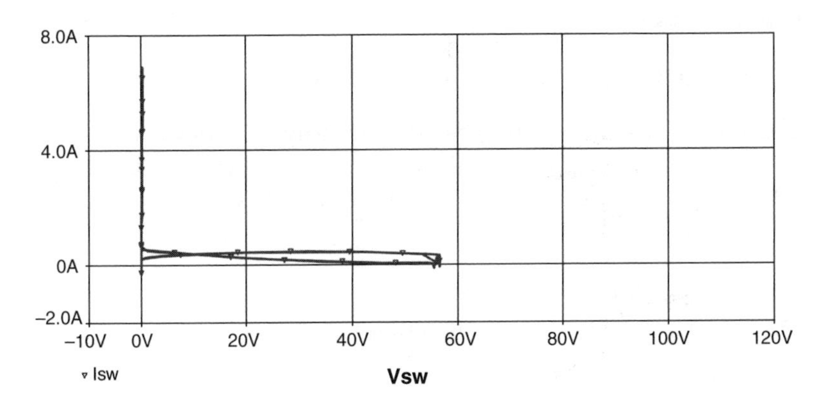

FIGURE 7.9
Switching trajectory of soft-switching buck converter.

the soft-switching technique effectively limits the conducted and radiated EMI emission.

7.3.4 Designing Rule to Minimize EMI/EMC

In order to reduce the EMI in certain stage and keep the reasonable EMC, some rules have to be considered during designing a DC/DC converter:

- Take soft-switching techniques ZCS, ZVS, and ZT
- Reduce the switching frequency as low as possible
- Reduce the working power
- Use fewer inductors
- House DC/DC converter in a shielded enclosure

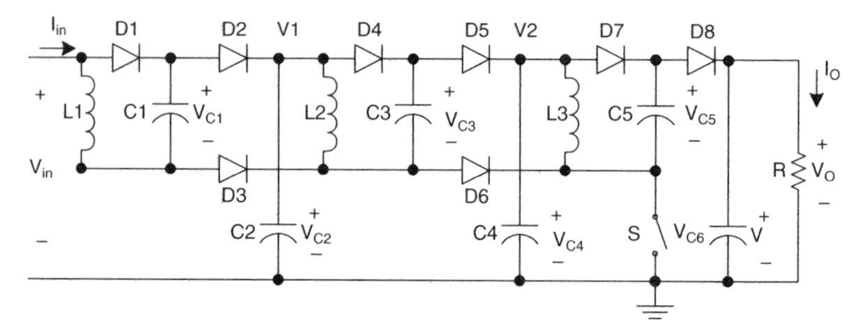

FIGURE 7.10
The 5000 V insulation test bench.

7.4 Some DC/DC Converter Applications

DC/DC conversion technique has been rapidly developed and has been widely applied in industrial applications and communication equipment. Some particular examples of DC/DC converters are presented here to demonstrate DC/DC converter application.

- A 5000 V insulation test bench
- MIT 42 V/14 V 3 kW dual direction DC/DC converter
- IBM 1.8 V/200 A power supply

7.4.1 A 5000 V Insulation Test Bench

Insulation test bench is the necessary equipment for semiconductor manufacturing organizations. An adjustable DC voltage power supply is the heart of this equipment. Traditional method to obtain the adjustable high DC voltage is a diode rectifier via a setting-up transformer. It is costly and large size with poor efficiency.

Using a positive output Luo-converter quadruple-lift circuit plus a general IC-chip TL494, can easily implement the high output voltage (say, 36 V to 1000 V) from a 24 V source. If higher voltage is required, it is available to implement 192 V to 5184 V, via a positive output super-lift Luo-converter triple-lift circuit, this diagram is shown in Figure 7.10. This circuit is small, effective, and low cost. The output voltage can be determined by

$$V_O = (\frac{2-k}{1-k})^3 V_{in} \tag{7.6}$$

TABLE 7.1

The Experimental Results of the 5000 V Test Bench

Conduction duty, k	0	0.1	0.2	0.3	0.4	0.5	0.6	0.7	0.8	0.82
Output voltage V_O (V)	192	226	273	244	455	648	1029	1953	5184	6760

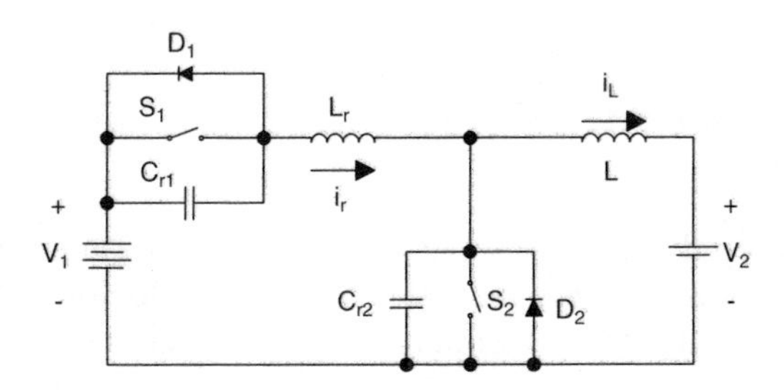

FIGURE 7.11
The MIT 42 V/14 V 3 kW dual direction DC/DC converter.

The conduction duty cycle k is only adjusted in the range 0 to 0.8 to carry out the output voltage in the range of 192 V to 5184 V. The experimental results are listed in Table 7.1. The measured data verified the advantage of this power supply.

7.4.2 MIT 42/14 V 3 KW DC/DC Converter

MIT 42/14 V 3 KW DC/DC converter was required to transfer 3 kW energy between two battery sources with 42 V and 14 V. The circuit diagram is shown in Figure 7.11. This is a two-quadrant zero-voltage-switching (ZVS) quasi-resonant-converter (QRC). The current in low-voltage side can be up to 250 A. This is a typical low-voltage strong-current converter. It is easier to carry out by ZVS-QRC.

This converter consists of two sources V_1 and V_2, one main inductor L, two main switches S_1 and S_2, two reverse-paralleled diodes D_1 and D_2, one resonant inductor L_r and two resonant capacitors C_{r1} and C_{r2}. The working conditions selected:

$$V_1 = 42\text{ V} \qquad V_2 = 14\text{ V}$$

$$\mathrm{L} = 470\ \mu\mathrm{H} \qquad \mathrm{C_{r1}} = \mathrm{C_{r2}} = \mathrm{C_r} = 1\ \mu\mathrm{F}$$

$$L_r = \begin{cases} 1\ \mu H & normal-operation \\ 9\ \mu H & low-current-operation \end{cases}$$

Therefore,

$$\omega_O = \frac{1}{\sqrt{L_r C_r}} = 10^6 \quad \text{rad/s} \tag{7.7}$$

$$Z_O = \sqrt{\frac{L_r}{C_r}} = 1 \quad \Omega \text{ (Normal operation)} \tag{7.8}$$

$$\alpha = \sin^{-1} \frac{V_1}{Z_O I_2} \tag{7.9}$$

It is easy to keep the quasi-resonance when the working current $I_2 > 50$ A. If the working current is too low, the resonant inductor will take a large value to guarantee the quasi-resonance state. This converter performs two-quadrant operation:

- Mode A (quadrant I) — Energy transferred from V_1 side to V_2 side
- Mode B (quadrant II) — Energy transferred from V_2 side to V_1 side

Assuming the working current is $I_2 = 100$ A and the converter works in mode A, following calculations are obtained:

$$\omega_O = \frac{1}{\sqrt{L_r C_r}} = 10^6 \text{ rad/s} \tag{7.10}$$

$$Z_O = \sqrt{\frac{L_r}{C_r}} = 1\ \Omega \tag{7.11}$$

$$\alpha = \sin^{-1} \frac{V_1}{Z_O I_2} = 24.83^\circ \tag{7.12}$$

$$t_1 = \frac{V_1 C_r}{I_2} = 0.42\ \mu\text{s} \tag{7.13}$$

$$t_2 = \frac{\pi + \alpha}{\omega_0} = 3.58\ \mu\text{s} \tag{7.14}$$

$$t_3 = \frac{1 + \cos\alpha}{V_1} I_2 L_r = \frac{1 + 0.908}{42} 100 * 10^{-6} = 4.54\ \mu\text{s} \tag{7.15}$$

$$t_4 = \frac{t_1 + t_2 + t_3}{V_1 / V_2 - 1} = \frac{0.42 + 3.58 + 4.54}{2} = 4.27\ \mu\text{s} \tag{7.16}$$

TABLE 7.2

The Experimental Test Results of MIT 42V/14 Converter

Mode	f (KHz)	I_1 (A)	I_2 (A)	I_L (A)	P_1 (W)	P_2 (W)	η (%)	PD (W/in.^{3})
A	78	77.1	220	220	3239	3080	95.1	23.40
A	80	78.3	220	220	3287	3080	93.7	23.58
A	82	81	220	220	3403	3080	90.5	24.01
B	68	220	69.9	220	3080	2939	95.3	22.28
B	70	220	68.3	220	3080	2871	93.2	22.04
B	72	220	66.6	220	3080	2797	90.8	21.77

Note: With the condition: $L_r = 1\ \mu H$, $C_{r1} = C_{r2} = 1\ \mu F$.

$$T = t_1 + t_2 + t_3 + t_4 = 0.42 + 3.58 + 4.54 + 4.27 = 12.81\ \mu s \tag{7.17}$$

$$f = \frac{1}{T} = \frac{1}{12.81} = 78.06\ \text{KHz} \tag{7.18}$$

$$k = \frac{t_3 + t_4}{T} = \frac{4.54 + 4.27}{12.81} = 0.688 \tag{7.19}$$

The volume of this converter is 270 cubic inches. The experimental test results in full power 3 kW are listed in Table 7.2. From the tested data, a high power density 22.85 W/in.^{3} and a high efficiency 93% are obtained. Because of soft-switching operation, the EMI is low and EMS and EMC are reasonable.

7.4.3 IBM 1.8 V/200 A Power Supply

This equipment is suitable for IBM next-generation computers with power supplies of 1.8 V/200 A. This is a ZCS SR DC/DC Luo-converter, and is shown in Figure 7.12. This converter is based on the double-current synchronous rectifier DC/DC converter plus zero-current-switching technique. It employs a hixaploid-core flat-transformer with the turn ratio $N = 1/12$. It has a six-unit ZCS synchronous rectifier double-current DC/DC converter. The six primary coils are connected in series, and six secondary circuits are connected in parallel. Each unit has particular input voltage V_{in} to be about 33 V, and can offer 1.8 V/35 A individually. Total output current is 210 A. The equivalent primary full current is $I_1 = 17.5$ A and equivalent primary load voltage is $V_2 = 130$ V. The ZCS natural resonant frequency is

$$\omega_O = \frac{1}{\sqrt{L_r C_r}} \tag{7.20}$$

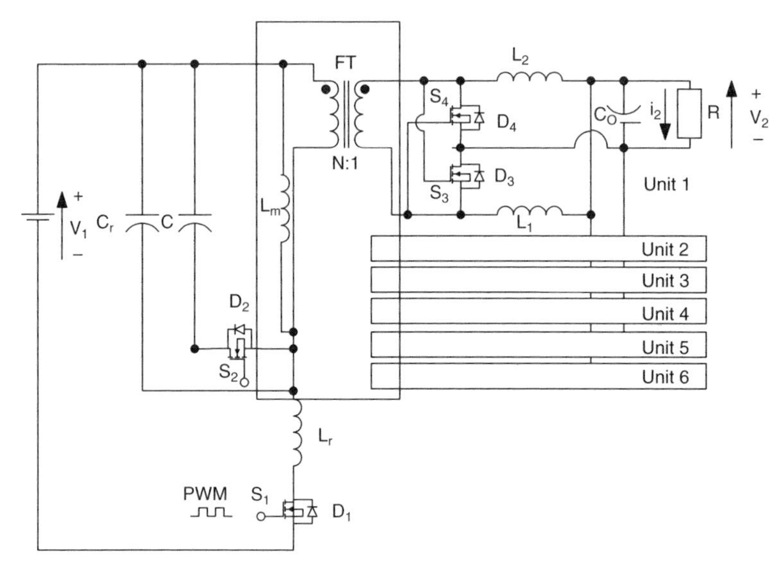

FIGURE 7.12
The IBM 1.8 V/200 A power supply.

$$Z_O = \sqrt{\frac{L_r}{C_r}} \tag{7.21}$$

$$\alpha = \sin^{-1}\frac{Z_O I_1}{V_1} \tag{7.22}$$

The main power supply is from public utility board (PUB) via a diode rectifier. Therefore V_1 is nearly 200 V, and the each unit input voltage V_{in} is about 33 V. Other calculation formulae are

$$t_1 = \frac{I_1 L_r}{V_1} \tag{7.23}$$

$$t_2 = \frac{\pi + \alpha}{\omega_0} \tag{7.24}$$

$$t_3 = \frac{1+\cos\alpha}{I_1} V_1 C_r \tag{7.25}$$

$$t_4 = \frac{V_1(t_1+t_2)}{I_1 V_2}\left(I_1 + \frac{V}{Z_0}\frac{\cos\alpha}{\pi/2+\alpha}\right) - (t_1+t_2+t_3) \tag{7.26}$$

$$T = t_1 + t_2 + t_3 + t_4 \tag{7.27}$$

$$f = \frac{1}{T} \tag{7.28}$$

$$k = \frac{t_1 + t_2}{T} \tag{7.29}$$

Real output voltage and input current are

$$V_O = kNV_1 - (R_L + R_S + \frac{L_m}{T} N^2) I_O \tag{7.30}$$

$$I_{in} = kNI_O \tag{7.31}$$

The power transfer efficiency is

$$\eta = \frac{V_O I_O}{V_{in} I_{in}} = 1 - \frac{R_L + R_S + \frac{L_m}{T} N^2}{kNV_{in}} I_O \tag{7.32}$$

The commercial unit of this power supply works in voltage closed-loop control with inner current closed-loop to keep the output voltage constant. Applying frequency is arranged in the band of 200 kHz to 250 kHz. The volume of the power supply is 14 cubic inches. The transfer efficiency is about 88 to 92% and power density is about 25.7 W/in.^{3}.

Bibliography

Bech, M., Blaabjerg, F., and Pederson, J., Random modulation techniques with fixed switching frequency for three-phase power converters, *IEEE Transactions on PE*, 15, 753, 2000.

Bogart, T.F., Jr., *Linear Electronics*, Maxwell Macmillan International, New York, 1994.

Chung, H., Hui, S., and Tse, K., Reduction of power converter EMI emission using soft-switching technique, *IEEE Trans on Electromagnetic Compatibility*, 40, 282, 1998.

Consoli, A., Musumeci, S., Oriti, G., and Testa, A., An innovative EMI reduction technique in power converters, *IEEE Transactions on Electromagnetic Compatibility*, 38, 567, 1996.

Heerema, M.D., *Designing for Electromagnetic Compatibility*, Proceedings of Hewlett-Packard Seminar, 1996.

Horowitz, P. and Hill, W., *The Art of Electronics*, 2nd ed., Cambridge, 1990.

Luo, F.L., Luo-resonator — a PWM signal generator applied to all DC/DC converters, *Power Supply Technologies and Applications*, Xi'an, China, 3, 198, 2000.

Luo, F.L. and Ye, H., Investigation of EMI, EMS, and EMC in power DC/DC converters, in *Proceedings of IEEE-PEDS'03*, Singapore, 2003, p. 628.
Tihanyi, L., *Electromagnetic Compatibility in Power Electronics*, J. K. Eckert & Company, Inc., Sarasota, FL, 1995.
Tse, K., Chung, H., and Hui, S., A comparative study of carrier-frequency modelation techniques for conducted EMI suppression in PWM converters, *IEEE Transactions on IE*, 49, 618, 2002.
Wait, J.V., Huelsman, L.P., and Korn, G.A., *Introduction to Operational Amplifier Theory and Applications*, 2nd ed., McGraw-Hill Inc., New York, 1992.

8

Energy Factor (EF) and Mathematical Modeling for Power DC/DC Converters

We have investigated the characteristics of all power DC/DC converters in steady-state in previous chapters. We will investigate the transient process characteristics of all power DC/DC converters in this chapter.

Power DC/DC converters have pumping-filtering process, and resonant process or voltage-lift operation or both. These circuits consist of several energy-storage elements. They are like an energy container that stores certain energy during performance. The stored energy will vary if the working conditions change. For example, once the power supply is on, the output voltage starts from zero since the energy-container is not filled. The transient process from one steady state to another depends on the pumping energy and stored energy. The same reason effects the interference discovery process since the stored energy usually has the inertia to effect the impulse response.

All switching power circuits work under the switching condition with high frequency f. It is thoroughly different from traditional continuous work condition. The obvious technical feature is that all parameters perform in a period $T = 1/f$, then gradually change period-by-period. The switching period T is the clue to investigate all switching power circuits. Catching the clue, we can define many brand new concepts (parameters) to describe the characteristics of switching power circuits. These new factors fill in the blanks of the knowledge in power electronics and conversion technology. We will carefully discuss the new concepts and their applications in this chapter.

8.1 Introduction

Four important factors — power factor (**PF**), power transfer efficiency (**η**), total harmonic distortion (**THD**) and ripple factor (**RF**) — well describe the characteristics of power systems. Unfortunately, all these factors are not available to use to describe the characteristics of power DC/DC converters and other high frequency-switching circuits.

Power DC/DC converters are equipped with a DC power supply source, pump-circuit, filter, and load. The load can be any type of the load, but most investigation is concerned with resistive load R and back EMF or battery. It means that the input and output voltages are nearly pure DC voltages with a very small ripple, e.g., the output voltage variation ratio is usually less than 1%. In this case, the corresponding *RF* is less than 0.001, which is always ignored.

Since all power is real power without reactive power jQ, we cannot use the power factor *PF* to describe the energy transferring process.

Because only the DC component exists without harmonics in input and output voltage, *THD* is not available to use to describe the energy transferring process and waveform distortion.

To simplify the research and analysis, we usually assume the condition without power losses during the power transferring process when investigating power DC/DC converters. Consequently, we assume the efficiency $\eta = 1 = 100\%$ for most descriptions of power DC/DC investigations. Otherwise, efficiency η must be considered for special investigations regarding the power losses.

In general conditions, all four factors are not available to apply to the analysis of power DC/DC converters. This situation left the designers of power DC/DC converters confused for a very long time. They would like to find other new parameters to describe the characteristics of power DC/DC converters.

Energy storage in power DC/DC converters has received attention for a long time. Unfortunately, as yet there is no clear concept to describe the phenomena and reveal the relationship between the stored energy and the characteristics of power DC/DC converters. Theoretically, we have defined a new concept — energy factor (EF) — and researched the relationship between EF and the mathematical modelling of power DC/DC converters. EF is a new concept in power electronics and conversion technology, and thoroughly differs from the traditional concepts, such as power factor (PF), power transfer efficiency (η), total harmonic distortion (THD), and ripple factor (RF). EF and the sub-sequential parameters can illustrate the system stability, reference response, and interference recovery. This investigation is helpful for system design and foreseeing DC/DC converters' characteristics.

Assume that the instantaneous input voltage and current of a DC/DC converter are $v_1(t)$ and $i_1(t)$, and their average values are V_1 and I_1. The instantaneous output voltage and current of a DC/DC converter are $v_2(t)$ and $i_2(t)$, and their average values are V_2 and I_2. The switching frequency is f, the switching period is $T = 1/f$, the conduction duty cycle is k, and the voltage transfer gain is $M = V_2/V_1$.

8.2 Pumping Energy (PE)

All power DC/DC converters have a pumping circuit to transfer the energy from the source to some energy-storage passive elements, e.g., inductors and capacitors. The pumping energy (**PE**) counts the input energy in a switching period *T*. Its calculation formula is:

$$PE = \int_0^T P_{in}(t)dt = \int_0^T V_1 i_i dt = V_1 I_1 T \qquad (8.1)$$

where

$$I_1 = \int_0^T i_1(t)dt$$

is the average value of the input current if the input voltage V_1 is constant. Usually the input average current I_1 depends on the conduction duty cycle.

8.2.1 Energy Quantization

In switching power circuits, the energy is not continuously flowing from source to actuator. The energy delivered in a switching period *T* from source to actuator is likely an energy quantum. Its value is the PE.

8.2.2 Energy Quantization Function

Equation (8.1) demonstrates that the energy quantum (PE) is the function of switching frequency f or period *T*, conduction duty cycle k, input voltage v_1, and current i_1. Since the variables T, k, v_1, and i_1 can vary with time, PE is the time-function. Usually, in a steady state the variables T, k, v_1, and i_1 cannot vary; consequently PE is a constant value in a steady state.

8.3 Stored Energy (SE)

Energy storage in power DC/DC converters has received attention long ago. Unfortunately, there is no clear concept to describe the phenomena and reveal the relationship between the stored energy and the characteristics of power DC/DC converters.

8.3.1 Stored Energy in Continuous Conduction Mode (CCM)

If a power DC/DC converter works in the continuous conduction mode, all components' voltages and currents are continuous.

8.3.1.1 Stored Energy (SE)

The stored energy (SE) in an inductor is

$$W_L = \frac{1}{2} L I_L^2 \tag{8.2}$$

The stored energy across a capacitor is

$$W_C = \frac{1}{2} C V_C^2 \tag{8.3}$$

Therefore, if there are n_L inductors and n_C capacitors, the total stored energy in a DC/DC converter is

$$SE = \sum_{j=1}^{n_L} W_{Lj} + \sum_{j=1}^{n_C} W_{Cj} \tag{8.4}$$

Usually, the stored energy is independent from the switching frequency f (as well as the switching period T). Since the inductor currents and the capacitor voltages rely on the conduction duty cycle k, the stored energy is dependent on the conduction duty cycle k. In subsequent descriptions, we call it stored energy.

8.3.1.2 Capacitor-Inductor Stored Energy Ratio (CIR)

Most power DC/DC converters consist of inductors and capacitors. Therefore, we can define the capacitor-inductor stored energy ratio (***CIR***).

$$CIR = \frac{\sum_{j=1}^{n_C} W_{Cj}}{\sum_{j=1}^{n_L} W_{Lj}} \tag{8.5}$$

8.3.1.3 Energy Losses (EL)

Usually, most analysis applied in DC/DC converters assumes no power losses, i.e., the input power is equal to the output power, $P_{in} = P_o$ or $V_1I_1 = V_2I_2$, so that pumping energy is equal to output energy in a period $PE = V_1I_1T = V_2I_2T$. It corresponds to the efficiency $\eta = V_2I_2T/PE = 100\%$.

Particularly, power losses always exist during the conversion process. They are caused by the resistance of the connection cables, resistance of the inductor and capacitor wire, and power losses across the semiconductor devices (diode, IGBT, MOSFET, and so on). We can sort them as the resistance power losses P_r, passive element power losses P_e, and device power losses P_d. The total power losses are P_{loss}.

$$P_{loss} = P_r + P_e + P_d$$

and

$$P_{in} = P_o + P_{loss} = P_o + P_e + P_e + P_d = V_2I_2 + P_e + P_e + P_d$$

Therefore,

$$EL = P_{loss} \times T = (P_r + P_e + P_d)T$$

The energy losses (EL) is in a period T,

$$EL = \int_0^T P_{loss} dt = P_{loss} T \tag{8.6}$$

Since the output energy in a period T is $(PE - EL) \times T$, we can define the efficiency η to be

$$\eta = \frac{P_o}{P_{in}} = \frac{P_{in} - P_{loss}}{P_{in}} = \frac{PE - EL}{PE} \tag{8.7}$$

If there is some energy loss $EL > 0$, the efficiency η is smaller than unity. If there is no energy loss during conversion process ($EL = 0$), the efficiency $\eta = 1$.

8.3.1.4 Stored Energy Variation on Inductors and Capacitors (VE)

The current flowing through an inductor has variation (ripple) Δi_L; the variation of stored energy in an inductor is

$$\Delta W_L = \frac{1}{2} L\left(I_{max}^2 - I_{min}^2\right) = LI_L \Delta i_L \tag{8.8}$$

where

$$I_{max} = I_L + \Delta i_L/2 \quad \text{and} \quad I_{min} = I_L - \Delta i_L/2$$

The voltage across a capacitor has variation (ripple) Δv_C; the variation of stored energy across a capacitor is

$$\Delta W_C = \frac{1}{2} C\left(V_{max}^2 - V_{min}^2\right) = CV_C \Delta v_C \tag{8.9}$$

where

$$V_{max} = V_C + \Delta v_C/2 \quad \text{and} \quad V_{min} = V_C - \Delta v_C/2$$

In the steady state of CCM, the total variation of the stored energy (VE) is

$$VE = \sum_{j=1}^{n_L} \Delta W_{Lj} + \sum_{j=1}^{n_C} \Delta W_{Cj} \tag{8.10}$$

8.3.2 Stored Energy in Discontinuous Conduction Mode (DCM)

If a power DC/DC converter works in the discontinuous conduction mode (DCM), some components' voltages and currents are discontinuous. In the steady state of the discontinuous conduction situation, some minimum currents through inductors and/or some minimum voltages across capacitors become zero. We define the filling coefficients m_L and m_C to describe the performance in DCM.

Usually, if the switching frequency f is high enough, the inductor's current is a triangle waveform. It increases and reaches I_{max} during the switching-on period kT, and decreases and reaches I_{min} during the switching-off period $(1 - k)T$. If it becomes zero at $t = t_1$ before the next switching-on, we consider that the converter is working in DCM. The waveform of the inductor's current is shown in Figure 8.1. The time t_1 should range $kT < t_1 < T$, and the filling coefficient m_L is

$$m_L = \frac{t_1 - kT}{(1 - k)T} \tag{8.11}$$

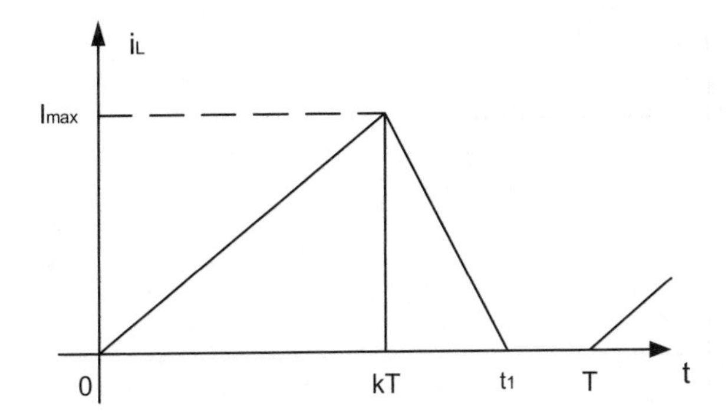

FIGURE 8.1
Discontinuous inductor current

where $0 < m_L < 1$. It means that the inductor's current only can fill the time period $m_L(1 - k)T$ during switch-off. In this case, I_{min} is equal to zero and the average current I_L is

$$I_L = I_{max}[m_L + (1 - m_L)k]/2 \tag{8.12}$$

and

$$\Delta i_L = I_{max} \tag{8.13}$$

Therefore,

$$\Delta W_L = LI_L\Delta i_L = LI^2_{max}\,[m_L + (1 - m_L)k]/2 \tag{8.14}$$

Analogously, we define the filling coefficient m_C to describe the capacitor voltage discontinuity. The waveform is shown in Figure 8.2. Time t_2 should be $kT < t_2 < T$, and the filling coefficient m_C is

$$m_C = \frac{t_2 - kT}{(1-k)T} \tag{8.15}$$

where $0 < m_C < 1$. It means that the capacitor's voltage only can fill the time period $m_C(1 - k)T$ during switch-off. In this case, V_{min} is equal to zero and the average voltage V_C is

$$V_c = V_{max}\,[m_C + (1 - m_C)k]/2 \tag{8.16}$$

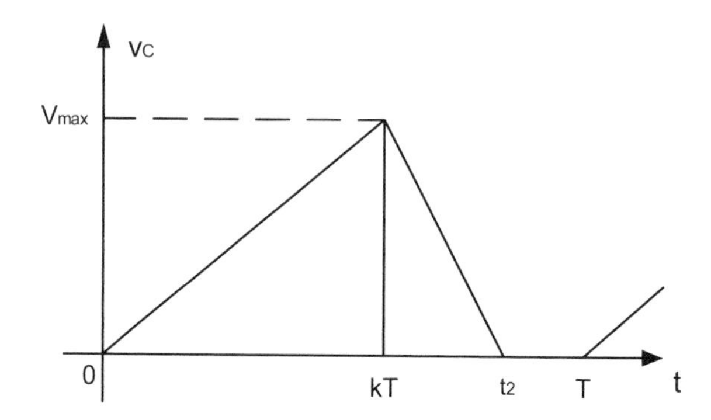

FIGURE 8.2
Discontinuous capacitor voltage.

and

$$\Delta v_C = V_{max} \tag{8.17}$$

Therefore,

$$\Delta W_C = CV_C \Delta v_C = CV_{max}^2 \,[m_C + (1 - m_C)k]/2 \tag{8.18}$$

When we consider that a converter works in DCM, it usually means that only one or two energy-storage elements, voltage and current, are discontinuous, not all elements. We use the parameter VE_D to present the total variation of the stored energy.

$$VE_D = \sum_{j=1}^{n_{L-d}} \Delta W_{Lj} + \sum_{j=n_{L-d}+1}^{n_L} \Delta W_{Lj} + \sum_{j=1}^{n_{C-d}} \Delta W_{Cj} + \sum_{j=n_{C-d}+1}^{n_C} \Delta W_{Cj} \tag{8.19}$$

where $n_{L\text{-}d}$ is the number of discontinuous inductor currents, and $n_{C\text{-}d}$ is the number of discontinuous capacitor voltages. We have discussed these cases in other papers. This formula is the same as equation (8.10). For convenience, rather than necessity, we use equation (8.10) to cover both CCM and CDM.

8.4 Energy Factor (EF)

As described in a previous section, the input energy in a period T is the pumping energy $PE = P_{in} \times T = V_{in} I_{in} \times T$. We now define the energy factor (EF) as the ratio of the stored energy (SE) over the pumping energy (PE):

$$EF = \frac{SE}{PE} = \frac{SE}{V_1 I_1 T} = \frac{\sum_{j=1}^{m} W_{Lj} + \sum_{j=1}^{n} W_{Cj}}{V_1 I_1 T} \tag{8.20}$$

Energy factor (*EF*) is a very important factor of a power DC/DC converter. It is usually independent from the conduction duty cycle k, and proportional to the switching frequency f (inversely proportional to the period *T*) since the pumping energy *PE* is proportional to the switching period *T*.

8.5 Variation Energy Factor (EF_V)

We also define the energy factor for the variation of stored energy (EF_V) that is the ratio of the variation of stored energy over the pumping energy:

$$EF_V = \frac{VE}{PE} = \frac{VE}{V_1 I_1 T} = \frac{\sum_{j=1}^{m} \Delta W_{Lj} + \sum_{j=1}^{n} \Delta W_{Cj}}{V_1 I_1 T} \tag{8.21}$$

Energy factor *EF* and variation energy factor EF_V are available to be used to describe the characteristics of power DC/DC converters. The applications are listed in next sections.

8.6 Time Constant τ and Damping Time Constant τ_d

We define the time constant τ and damping time constant τ_d of a power DC/DC converter in this section for the applications in the next section.

8.6.1 Time Constant τ

The **time constant** τ of a power DC/DC converter is a new concept to describe the transient process of a DC/DC converter. If there are no power losses in the converter, it is defined:

$$\tau = \frac{2T \times EF}{1 + CIR} \tag{8.22}$$

This time constant τ is independent from switching frequency f (or period $T = 1/f$). It is available to estimate the converter responses for a unit-step function and impulse interference.
If there are power losses and $\eta < 1$, it is defined:

$$\tau = \frac{2T \times EF}{1+CIR}\left(1+CIR\frac{1-\eta}{\eta}\right) \tag{8.23}$$

The time constant t still is independent from switching frequency f (or period $T = 1/f$) and conduction duty cycle k. If there is no power loss, $\eta = 1$, equation (8.23) becomes (8.22). Usually, the higher the power losses (the lower efficiency η), the larger the time constant τ since $CIR > 1$.

8.6.2 Damping Time Constant τ_d

The damping time constant τ_d of a power DC/DC converter is new concept to describe the transient process of a DC/DC converter. If there are no power losses it is defined:

$$\tau_d = \frac{2T \times EF}{1+CIR}CIR \tag{8.24}$$

This damping time constant τ_d is independent from switching frequency f (or period $T = 1/f$). It is available to estimate the oscillation responses for a unit-step function and impulse interference.
If there are power losses and $\eta < 1$, it is defined:

$$\tau_d = \frac{2T \times EF}{1+CIR}\frac{CIR}{\eta+CIR(1-\eta)} \tag{8.25}$$

The damping time constant τ_d is also independent from switching frequency f (or period $T = 1/f$) and conduction duty cycle k. If there is no power loss, $\eta = 1$, equation (8.25) becomes (8.24). Usually, the higher the power losses (the lower efficiency η), the smaller the damping time constant τ_d since CIR > 1.

8.6.3 Time Constants Ratio ξ

The **time constants ratio** ξ of a power DC/DC converter is a new concept to describe the transient process of a DC/DC converter. If there are no power losses, it is defined:

$$\xi = \frac{\tau_d}{\tau} = CIR \tag{8.26}$$

This time constant ratio is independent from switching frequency f (or period $T = 1/f$). It is available to estimate the oscillation responses for a unit-step function and impulse interference.
If there are power losses and $\eta < 1$, it is defined:

$$\xi = \frac{\tau_d}{\tau} = \frac{CIR}{\eta\left(1 + CIR\frac{1-\eta}{\eta}\right)^2} \tag{8.27}$$

The time constant ratio is still independent from switching frequency f (or period $T = 1/f$) and conduction duty cycle k. If there is no power loss, $\eta = 1$, equation (8.27) becomes (8.26). Usually, the higher the power losses (the lower efficiency η), the smaller the time constant ratio ξ since $CIR > 1$. From this analysis, most power DC/DC converters with lower power losses possess the output voltage oscillation when the converter operation state changes. And vice versa, power DC/DC converters with high power losses will possess the output voltage smoothening when the converter operation state changes.

By cybernetic theory, we can estimate the unit-step function response using the ratio ξ. If the ratio ξ is equal to or smaller than 0.25 the corresponding unit-step function response has no oscillation and overshot. Vice versa, if the ratio ξ is greater than 0.25, the corresponding unit-step function response has oscillation and overshot. The higher the value of ratio ξ, the heavier oscillation with higher overshot.

8.6.4 Mathematical Modeling for Power DC/DC Converters

The mathematical modeling for all power DC/DC converters is

$$G(s) = \frac{M}{1 + s\tau + s^2\tau\tau_d} \tag{8.28}$$

where M the voltage transfer gain: $M = V_O/V_{in}$,
τ the time constant in (8.23),
τ_d the damping time constant in (8.25), $\tau_d = \xi\tau$
s the Laplace operator in the s-domain.

Using this mathematical model of power DC/DC converters is significantly easier to describe the characteristics of power DC/DC converters. In order

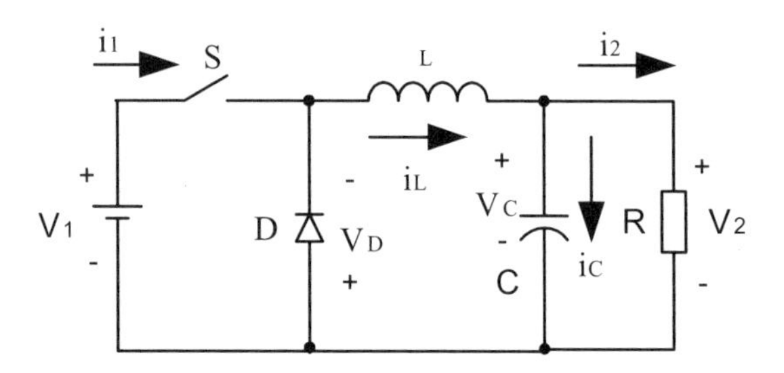

FIGURE 8.3
Buck converter.

to verify this theory, few converters are investigated to demonstrate the characteristics of power DC/DC converters and applications of the theory.

8.7 Examples of Applications

In order to demonstrate the parameters' calculation, some examples are presented in this section. A buck converter, super-lift Luo-converter, boost converter, buck-boost converter, and positive output Luo-converter are used for this purpose,

8.7.1 A Buck Converter in CCM

We will carefully discuss the mathematical model for buck converter in various conditions in this sub-section.

8.7.1.1 *Buck Converter without Energy Losses (r_L = 0 Ω)*

A buck converter shown in Figure 8.3 has the components values: $V_1 = 40$ *V*, *L* = 250 μ*H* with resistance $r_L = 0$ Ω, *C* = 60 μ*F*, *R* = 10 Ω, the switching frequency f = 20 kHz (*T* = 1/*f* = 50 μs) and conduction duty cycle *k* = 0.4. This converter is stable and works in CCM.

Therefore, we have got the voltage transfer gain $M = 0.4$, i.e. $V_2 = V_C = MV_1 = 0.4 \times 40 = 16$ *V*. $I_L = I_2 = 1.6$ *A*, $P_{loss} = 0$ *W* and $I_1 = 0.64$ *A*. The parameter *EF* and others are listed below:

$$PE = V_1 I_1 T = 40 * 0.64 * 50\mu = 1.28\,\text{mJ}, W_L = \frac{1}{2} L I_L^2 = 0.5 * 250\mu * 1.6^2 = 0.32\,\text{mJ},$$

$$W_C = \frac{1}{2}CV_C^2 = 0.5 * 60\mu * 16^2 = 7.68\,\text{mJ}, SE = W_L + W_C = 0.32 + 7.68 = 8\,\text{mJ},$$

$$EF = \frac{SE}{PE} = \frac{8}{1.28} = 6.25, \qquad CIR = \frac{W_C}{W_L} = \frac{7.68}{0.32} = 24$$

$$EL = P_{loss} * T = 0\,\text{mJ}, \qquad \eta = \frac{P_O}{P_O + P_{loss}} = 1.$$

$$\tau = \frac{2T * EF}{1 + CIR}\left(1 + CIR\frac{1-\eta}{\eta}\right) = 25 \quad \mu s, \ \tau_d = \frac{2T * EF}{1 + CIR}\frac{CIR}{\eta + CIR(1-\eta)} = 600\,\mu s$$

$$\xi = \frac{\tau_d}{\tau} = \frac{CIR}{\eta\left(1 + CIR\frac{1-\eta}{\eta}\right)^2} = 25 >> 0.25$$

By cybernetic theory, since the damping time constant τ_d is much larger than the time constant τ, the corresponding ratio ξ is 25 >> 0.25. The output voltage has heavy oscillation with high overshot. The corresponding transfer function is

$$G(s) = \frac{M}{1 + s\tau + s^2\tau\tau_d} = \frac{M / \tau\tau_d}{(s + s_1)(s + s_2)} \tag{8.29}$$

where $s_1 = \sigma + j\omega$ and $s_2 = \sigma - j\omega$

with

$$\sigma = \frac{1}{2\tau_d} = \frac{1}{1200\mu s} = 833\text{ Hz}$$

and

$$\omega = \frac{\sqrt{4\tau\tau_d - \tau^2}}{2\tau\tau_d} = \frac{\sqrt{60000 - 625}}{30000\mu} = \frac{243.67}{30000\mu} = 8122\text{ rad/s}$$

The unit-step function response is

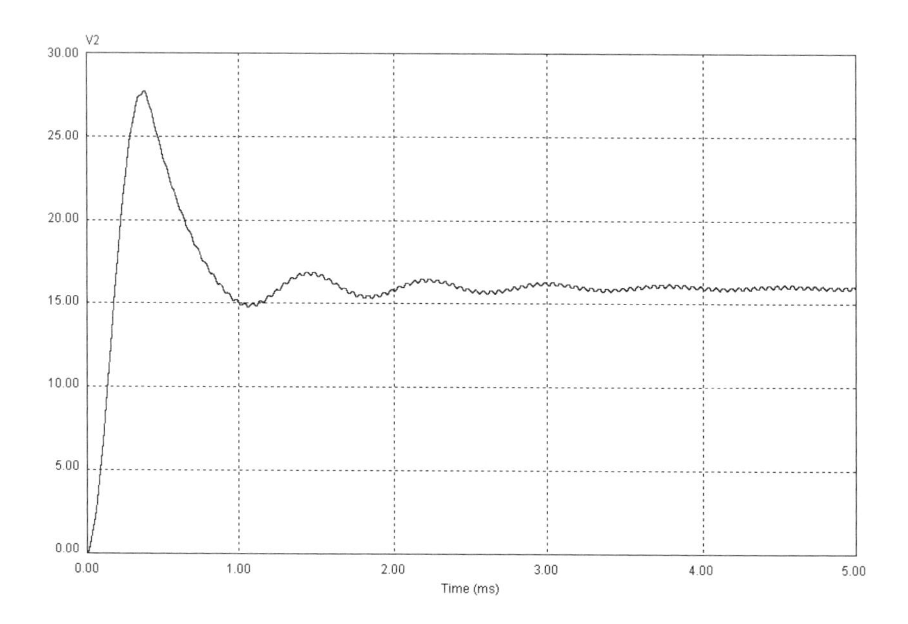

FIGURE 8.4
Buck converter unit-step function response.

$$v_2\left(t\right) = 16\left[1 - e^{\frac{t}{0.0012}}\left(\cos 122t - 0.1026\sin 8122t\right)\right]V$$

The unit-step function response (transient process) has oscillation progress with damping factor σ and frequency ω. The simulation result is shown in Figure 8.4.

The impulse interference response is

$$\Delta v_2\left(t\right) = 0.205Ue^{\frac{t}{0.0012}}\sin 8122t$$

where U is the interference signal. The impulse response (interference recovery process) has oscillation progress with damping factor σ and frequency ω. The simulation result is shown in Figure 8.5.

In order to verify the analysis, calculation, and simulation results; we constructed a test rig with the same conditions. The corresponding test results are shown in Figures 8.6 and 8.7.

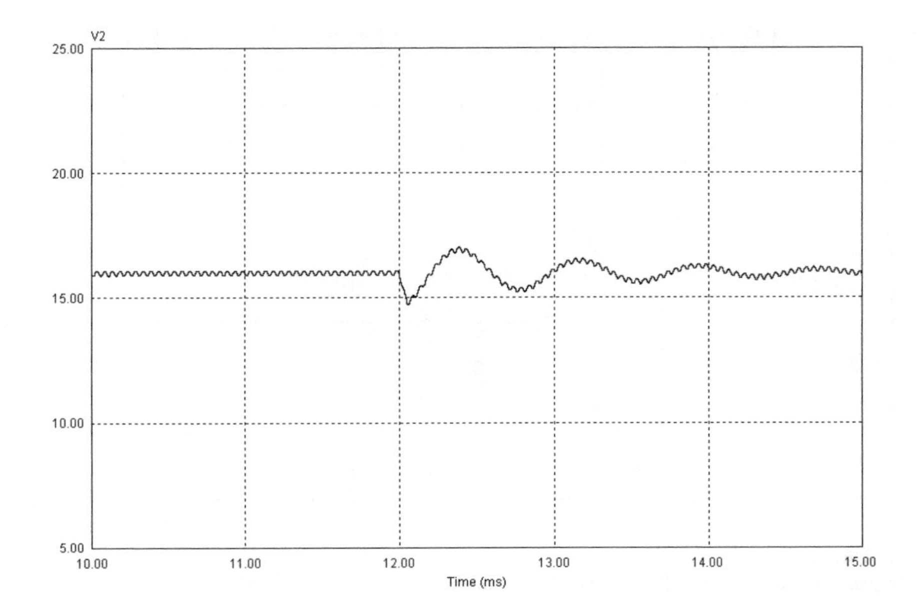

FIGURE 8.5
Buck converter impulse response.

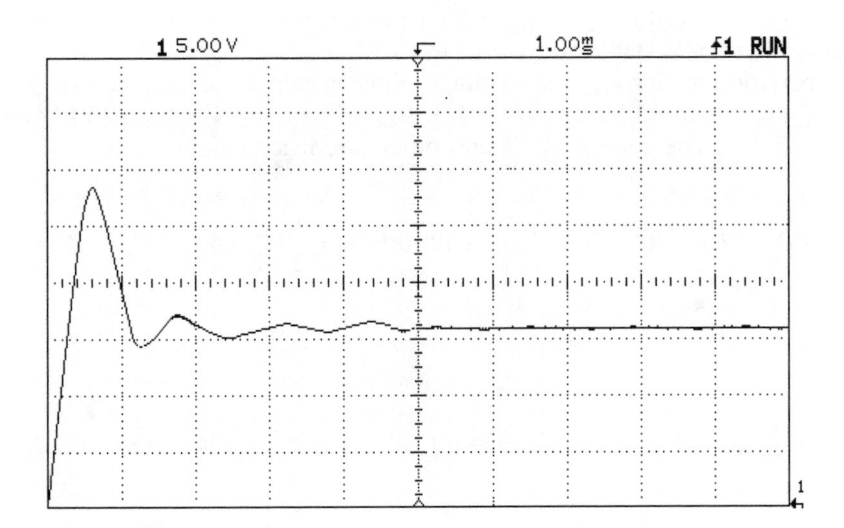

FIGURE 8.6
Unit-step function responses of buck converter (experiment).

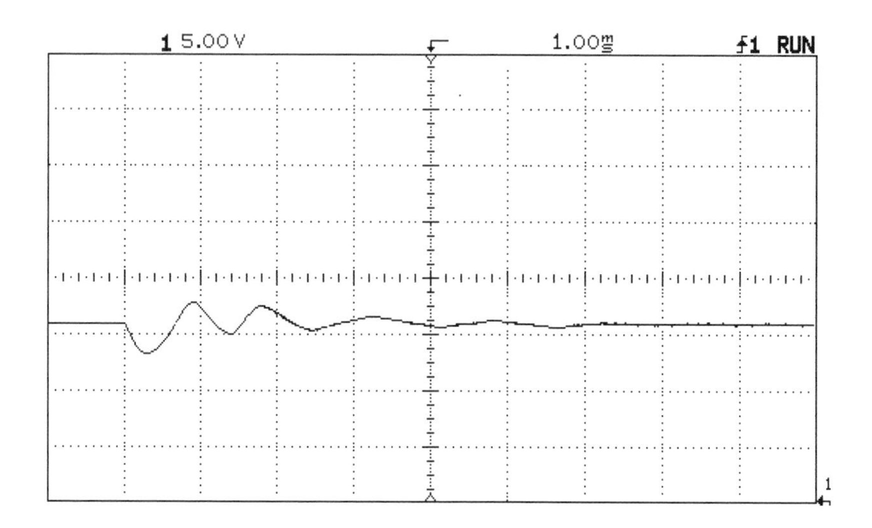

FIGURE 8.7
Impulse responses of buck converter (experiment).

8.7.1.2 Buck Converter with Small Energy Losses (r_L = 1.5 Ω)

A buck converter shown in Figure 8.3 has the components values: V_1 = 40 *V*, *L* = 250 μH with resistance r_L = 1.5 Ω, *C* = 60 mF, R = 10 Ω, the switching frequency f = 20 kHz ($T = 1/f$ = 50 μs) and conduction duty cycle k = 0.4. This converter is stable and works in CCM.

Therefore, we have got the voltage transfer gain M = 0.35, i.e. $V_2 = V_C = MV_1$ = 0.35 x 40 = 14 *V*. $I_L = I_2$ = 1.4 *A*, $P_{loss} = I^2{}_L \times r_L = 1.4^2 \times 1.5$ = 2.94 *W* and I_1 = 0.564 A. The parameter *EF* and others are listed below:

$$PE = V_1 I_1 T = 40 * 0.564 * 50\mu = 1.128\,\text{mJ}, W_L = \frac{1}{2} L I_L^2 = 0.5 * 250\mu * 1.4^2 = 0.245\,\text{mJ},$$

$$W_C = \frac{1}{2} C V_C^2 = 0.5 * 60\mu * 14^2 = 5.88\,\text{mJ}, SE = W_L + W_C = 0.245 + 5.88 = 6.125\,\text{mJ},$$

$$EF = \frac{SE}{PE} = \frac{6.125}{1.128} = 5.43 \qquad CIR = \frac{W_C}{W_L} = \frac{5.88}{0.2345} = 24$$

$$EL = P_{loss} * T = 2.94 * 50 = 0.147\,\text{mJ}, \qquad \eta = \frac{P_O}{P_O + P_{loss}} = 0.87,$$

$$\tau = \frac{2T * EF}{1+CIR}\left(1 + CIR\frac{1-\eta}{\eta}\right) = 99.6 \quad \mu s_{\tau_d} = \frac{2T * EF}{1+CIR}\frac{CIR}{\eta + CIR(1-\eta)} = 130.6\,\mu s$$

$$\xi = \frac{\tau_d}{\tau} = \frac{CIR}{\eta\left(1 + CIR\frac{1-\eta}{\eta}\right)^2} = 1.31 >> 0.25$$

By cybernetic theory, since the damping time constant τ_d is larger than the time constant τ, the corresponding ratio ξ is 1.31 >> 0.25. The output voltage has heavy oscillation with high overshot. The corresponding transfer function is

$$G(s) = \frac{M}{1 + s\tau + s^2\tau\tau_d} = \frac{M/\tau\tau_d}{(s+s_1)(s+s_2)} \tag{8.30}$$

where $s_1 = \sigma + j\omega$ and $s_2 = \sigma - j\omega$

with

$$\sigma = \frac{1}{2\tau_d} = \frac{1}{261.2\mu s} = 3833\ \text{Hz}$$

and

$$\omega = \frac{\sqrt{4\tau\tau_d - \tau^2}}{2\tau\tau_d} = \frac{\sqrt{52031 - 9920}}{26015.5} = \frac{205.2}{26015.5\mu} = 7888\ \text{rad/s}$$

The unit-step function response is

$$v_2(t) = 14\left[1 - e^{\frac{t}{0.000261}}\left(\cos 7888t - 0.486\sin 7888t\right)\right]V$$

The unit-step function response (transient process) has oscillation progress with damping factor σ and frequency ω. The simulation resuslt is shown in Figure 8.8.

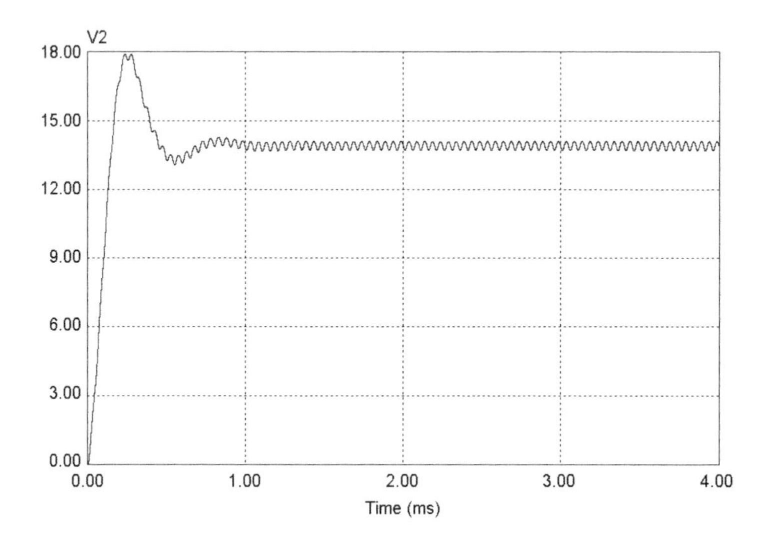

FIGURE 8.8
Buck converter unit-step function response (r_L = 1.5 Ω).

The impulse interference response is

$$\Delta v_2\left(t\right) = 0.975Ue^{\frac{t}{0.000261}}\sin 7888t$$

where U is the interference signal. The impulse response (interference recovery process) has oscillation progress with damping factor σ and frequency ω. The simulation result is shown in Figure 8.9.

In order to verify the analysis, calculation, and simulation results; we constructed a test rig with the same conditions. The corresponding test results are shown in Figures 8.10 and 8.11.

8.7.1.3 Buck Converter with Energy Losses (r_L = 4.5 Ω)

A Buck converter shown in Figure 8.3 has the components values: $V_1 = 40$ V, $L = 250$ μH with resistance $r_L = 4.5$ Ω, C = 60 μF, $R = 10$ Ω, the switching frequency f = 20 kHz ($T = 1/f = 50$ μs) and conduction duty cycle $k = 0.4$. This converter is stable and works in CCM.

Therefore, we have the voltage transfer gain $M = 0.2756$, i.e. $V_2 = V_C = MV_1 = 0.2756 \times 40 = 11$ V. $I_L = I_2 = 1.1$ A, $P_{loss} = I^2_L \times r_L = 1.1^2 \times 4.5 = 5.445$ W and $I_1 = 0.4386$ A. The parameter EF and others are listed below:

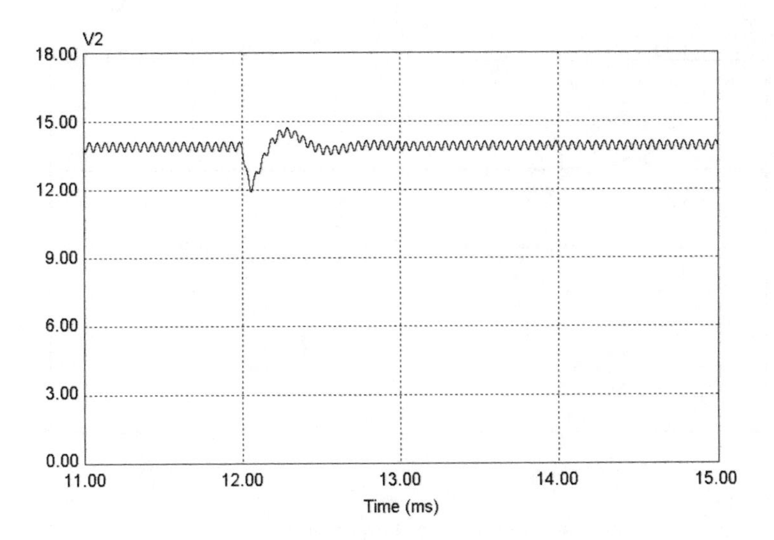

FIGURE 8.9
Buck converter impulse response (r_L = 1.5 Ω).

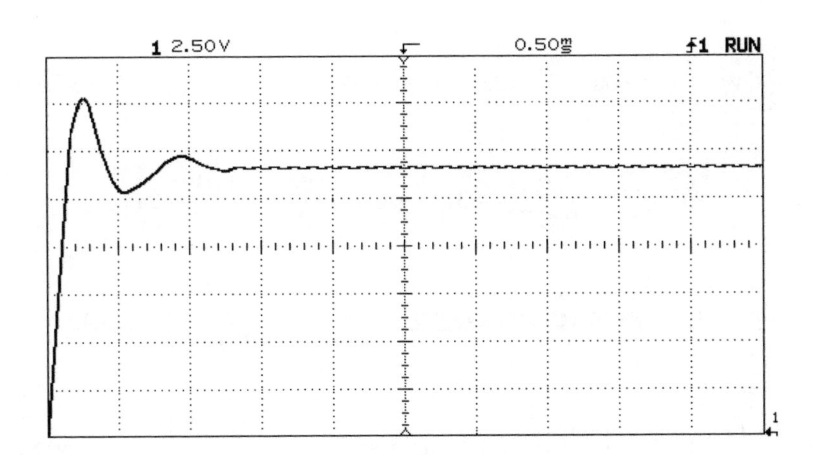

FIGURE 8.10
Unit-step function responses of buck converter (r_L = 1.5 Ω experiment).

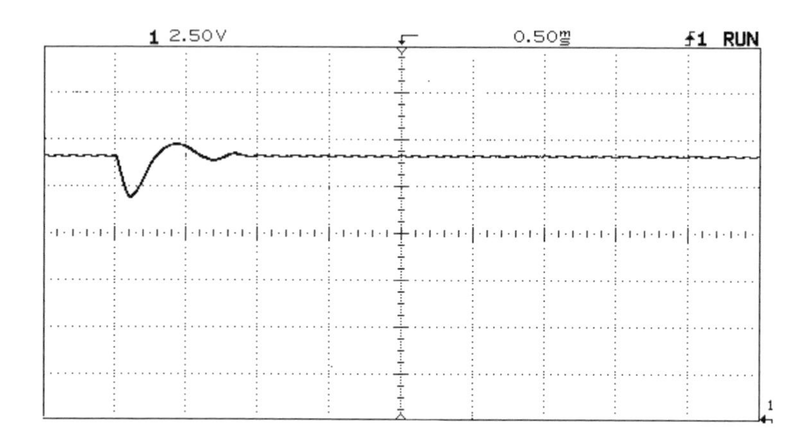

FIGURE 8.11
Impulse responses of buck converter (r_L = 1.5 Ω experiment).

$$PE = V_1 I_1 T = 40 * 0.4386 * 50\mu = 0.877 \text{ mJ}, W_L = \frac{1}{2} L I_L^2 = 0.5 * 250\mu * 1.1^2 = 0.151 \text{ mJ},$$

$$W_C = \frac{1}{2} C V_C^2 = 0.5 * 60\mu * 11^2 = 3.63 \text{ mJ}, SE = W_L + W_C = 0.151 + 3.63 = 3.781 \text{ mJ},$$

$$EF = \frac{SE}{PE} = \frac{3.781}{0.877} = 4.31, \qquad CIR = \frac{W_C}{W_L} = \frac{3.63}{0.151} = 24$$

$$EL = P_{loss} * T = 5.445 * 50 = 0.2722 \text{ mJ}, \qquad \eta = \frac{P_O}{P_O + P_{loss}} = 0.689,$$

$$\tau = \frac{2T * EF}{1 + CIR}\left(1 + CIR \frac{1-\eta}{\eta}\right) = 203.2 \quad \mu s, \ \tau_d = \frac{2T * EF}{1 + CIR} \frac{CIR}{\eta + CIR(1-\eta)} = 50.8 \, \mu s$$

$$\xi = \frac{\tau_d}{\tau} = \frac{CIR}{\eta\left(1 + CIR \frac{1-\eta}{\eta}\right)^2} = 0.25$$

By cybernetic theory, since the damping time constant τ_d is the critical value, the corresponding ratio ξ is equal to 0.25. The output voltage has no oscillation. The corresponding transfer function is

$$G(s) = \frac{M}{1 + s\tau + s^2\tau\tau_d} = \frac{M/\tau\tau_d}{(s+\sigma)^2} \tag{8.31}$$

where

$$\sigma = \frac{1}{2\tau_d} = \frac{1}{101.2\mu} = 9843 \text{ Hz}$$

The unit-step function response is

$$v_2(t) = 11\left[1 - \left(1 + \frac{t}{0.0001016}\right)e^{\frac{t}{0.0001016}}\right]V$$

The unit-step function response (transient process) has no oscillation progress with damping factor σ. The simulation result is shown in Figure 8.12. The impulse interference response is

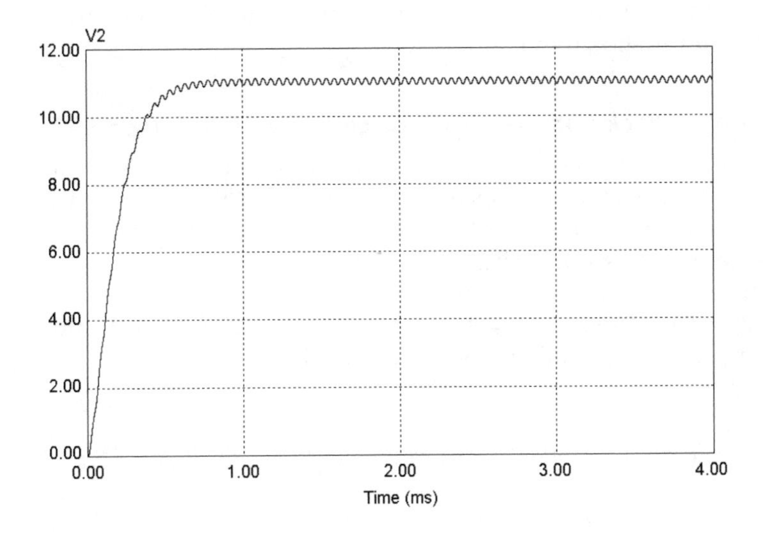

FIGURE 8.12
Buck converter unit-step function response (r_L = 4.5 Ω).

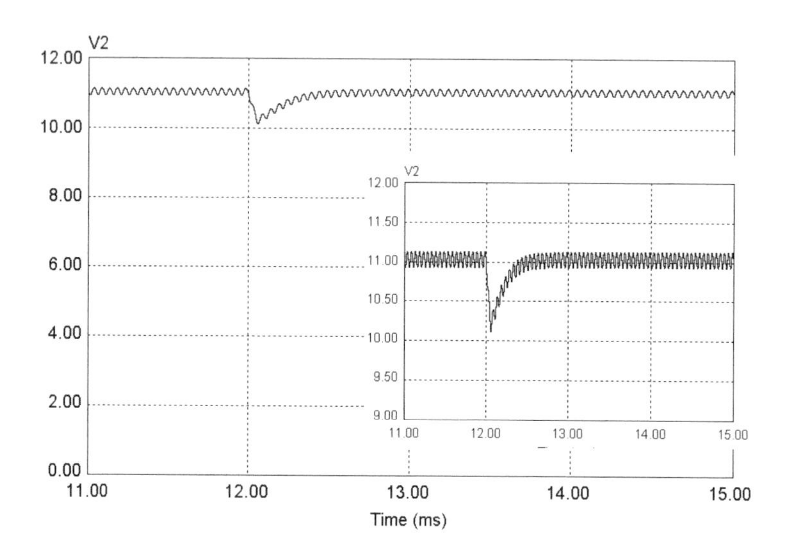

FIGURE 8.13
Buck converter Impulse response ($r_L = 4.5\ \Omega$).

$$\Delta v_2(t) = \frac{t}{0.0000508} U e^{\frac{t}{0.0001016}}$$

where U is the interference signal. The impulse response (interference recovery process) has no oscillation progress with damping factor σ. The simulation result is shown in Figure 8.13.

8.7.1.4 Buck Converter with Large Energy Losses ($r_L = 6\ \Omega$)

A buck converter shown in Figure 8.3 has the components values: $V_1 = 40$ V, $L = 250\ \mu$H with resistance $r_L = 6\ \Omega$, $C = 60\ \mu$F, $R = 10\ \Omega$, the switching frequency f = 20 kHz ($T = 1/f = 50\ \mu s$) and conduction duty cycle $k = 0.4$. This converter is stable and works in CCM.

Therefore, we have the voltage transfer gain $M = 0.25$, i.e. $V_2 = V_C = MV_1 = 0.25 \times 40 = 10\ V$. $I_L = I_2 = 1\ A$, $P_{loss} = I^2_L \times r_L = 1^2 \times 6 = 6\ W$ and $I_1 = 0.4\ A$. The parameter EF and others are listed below:

$$PE = V_1 I_1 T = 40 * 0.4 * 50\mu = 0.8\,\text{mJ}, W_L = \frac{1}{2} L I_L^2 = 0.5 * 250\mu * 1^2 =$$

$$0.125\,\text{mJ},$$

$$W_C = \frac{1}{2}CV_C^2 = 0.5*60\mu*10^2 = 3\,\text{mJ}, SE = W_L + W_C = 0.125+3 = 3.125\,\text{mJ},$$

$$EF = \frac{SE}{PE} = \frac{3.125}{0.8} = 3.9, \qquad CIR = \frac{W_C}{W_L} = \frac{3}{0.125} = 24$$

$$EL = P_{loss} * T = 6*50 = 0.3\,\text{mJ}, \qquad \eta = \frac{P_O}{P_O + P_{loss}} = 0.625,$$

$$\tau = \frac{2T*EF}{1+CIR}\left(1+CIR\frac{1-\eta}{\eta}\right) = 240.3 \quad \mu s,\ \tau_d = \frac{2T*EF}{1+CIR}\frac{CIR}{\eta+CIR(1-\eta)} = 38.9\,\mu s$$

$$\xi = \frac{\tau_d}{\tau} = \frac{CIR}{\eta\left(1+CIR\frac{1-\eta}{\eta}\right)^2} = 0.162 < 0.25$$

By cybernetic theory, since the damping time constant τ_d is smaller than the time constant τ, the corresponding ratio ξ is 0.162 < 0.25. The output voltage has no oscillation. The corresponding transfer function is

$$G(s) = \frac{M}{1+s\tau+s^2\tau\tau_d} = \frac{M/\tau\tau_d}{(s+\sigma_1)(s+\sigma_2)} \tag{8.32}$$

with

$$\sigma_1 = \frac{\tau+\sqrt{4\tau\tau_d-\tau^2}}{2\tau\tau_d} = \frac{240.3+142.66}{18695.3\mu} = 20500 \quad \text{Hz}$$

$$\sigma_2 = \frac{\tau-\sqrt{4\tau\tau_d-\tau^2}}{2\tau\tau_d} = \frac{240.3-142.66}{18695.3\mu} = 5200 \quad \text{Hz}$$

The unit-step function response is

$$v_2(t) = 10(1 + 0.342e^{-20500t} - 1.342^{e-5200t})\ V$$

The unit-step function response (transient process) has no oscillation progress with damping factor σ_2. The simulation result is shown in Figure 8.14.

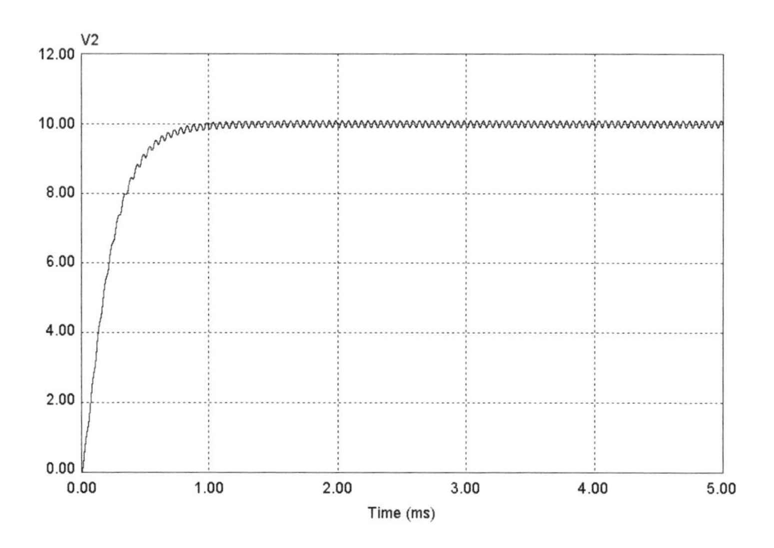

FIGURE 8.14
Buck converter unit-step function response ($r_L = 6\ \Omega$).

The impulse interference response is

$$\Delta v_2(t) = 1.684U(e^{-20500t} - e^{-5200t})$$

where U is the interference signal. The impulse response (interference recovery process) has no oscillation progress with damping factor σ_2. The simulation result is shown in Figure 8.15.

8.7.2 A Super-Lift Luo-Converter in CCM

Figure 8.16 shows a super-lift Luo-converter with the conduction duty k = 0.5. The components values are $V_1 = 20\ V$, f = 50 kHz ($T = 20\ \mu s$), $L = 100$ μH with resistance $r_L = 0.12\ \Omega$, $C_1 = 2500\ \mu F$, $C_2 = 800\ \mu F$ and R = 10 Ω. This converter is stable and works in CCM.

Therefore, we have the voltage transfer gain $M = 2.863$, i.e., the output voltage $V_2 = V_{C2} = 57.25\ V$. $V_{C1} = V_1 = 20\ V$, $I_1 = 17.175\ A$, $I_2 = 5.725$ A, $I_L = 11.45\ A$ and $P_{loss} = I^2_L \times r_L = 11.45^2 \times 0.12 = 15.73\ W$. The parameter EF and others are listed below:

$$PE = V_1 I_1 T = 20 * 17.175 * 20\mu = 6.87\,\text{mJ}, W_L = \frac{1}{2} L I_L^2 = 0.5 * 100\mu * 11.45^2 =$$

$$6.555\,\text{mJ},$$

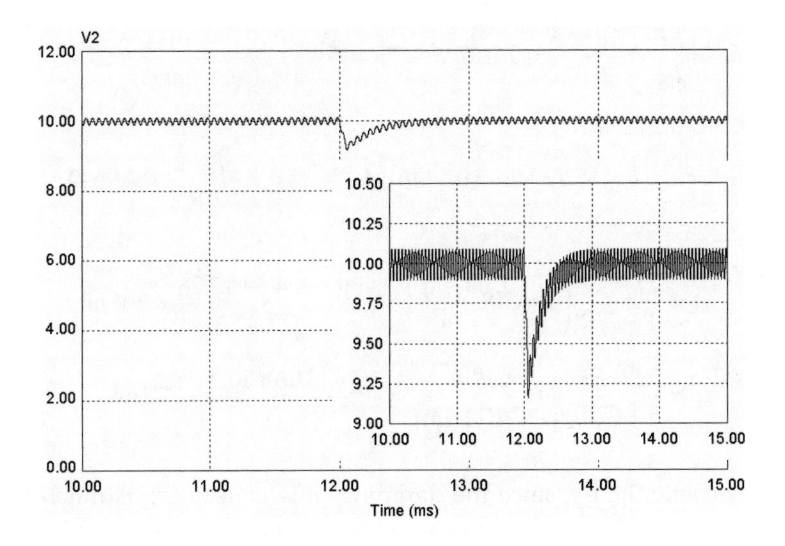

FIGURE 8.15
Buck converter impulse response (r_L = 6 Ω).

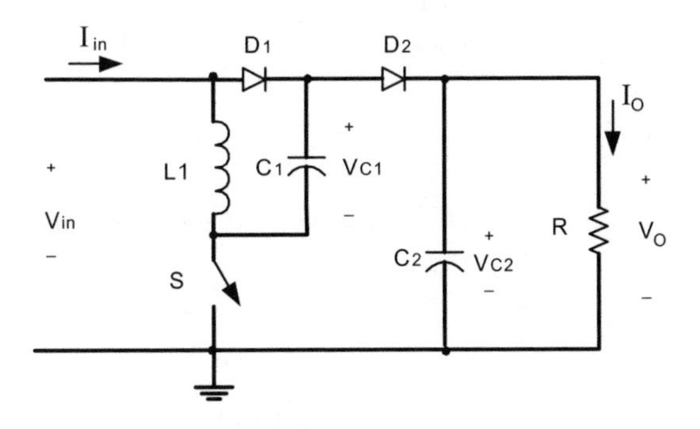

FIGURE 8.16
Super-lift Luo-converter.

$$W_{C1} = \frac{1}{2}C_1V_{C1}^2 = 0.5 * 2500\mu * 20^2 = 500\,\text{mJ},$$

$$W_{C2} = \frac{1}{2}C_2V_{C2}^2 = 0.5 * 800\mu * 57.25^2 = 1311\,\text{mJ},$$

$$SE = W_L + W_{C1} + W_{C2} = 6.555 + 500 + 1311 = 1817.6 \text{ mJ},$$

$$EF = \frac{SE}{PE} = \frac{1817.6}{6.87} = 264.6 \qquad CIR = \frac{W_{C1} + W_{c2}}{W_L} = \frac{1811}{6.555} = 276.3$$

$$EL = P_{loss}T = 15.73 * 20 = 0.3146\,\text{mJ}, \qquad \eta = \frac{P_O}{P_O + P_{loss}} = 0.9542,$$

$$\tau = \frac{2T \times EF}{1+CIR}\left(1 + CIR\frac{1-\eta}{\eta}\right) = \frac{40\mu * 264.6 * 13.26}{277.3} = 506\,\mu s$$

$$\tau_d = \frac{2T \times EF}{1+CIR}\frac{CIR}{\eta + CIR(1-\eta)} = \frac{40 * 264.6 * 20.3}{277.3} = 775\,\mu s$$

By cybernetic theory, since the damping time constant τ_d is much larger than the time constant τ, the corresponding ratio $\xi = 775/506 = 1.53 >> 0.25$. The output voltage has heavy oscillation with high overshot. The transfer function of this converter has two poles ($-s_1$ and $-s_2$) that are located in the left-hand half plane (LHHP).

$$G(s) = \frac{M}{1 + s\tau + s^2\tau\tau_d} = \frac{M/\tau\tau_d}{(s+s_1)(s+s_2)} \tag{8.33}$$

where

$$s_1 = \sigma + j\omega \quad \text{and} \quad s_2 = \sigma - j\omega$$

with

$$\sigma = \frac{1}{2\tau_d} = \frac{1}{1.55ms} = 645\ \text{Hz}$$

$$\omega = \frac{\sqrt{4\tau\tau_d - \tau^2}}{2\tau\tau_d} = \frac{\sqrt{1686400 - 295936}}{843200} = \frac{1197.2}{843200\mu} = 1398\ \text{rad/s}$$

The unit-step function response is

$$v_2(t) = 57.25\left[1 - e^{\frac{t}{0.00155}}\left(\cos 1398t - 0.461 \sin 1398t\right)\right]V$$

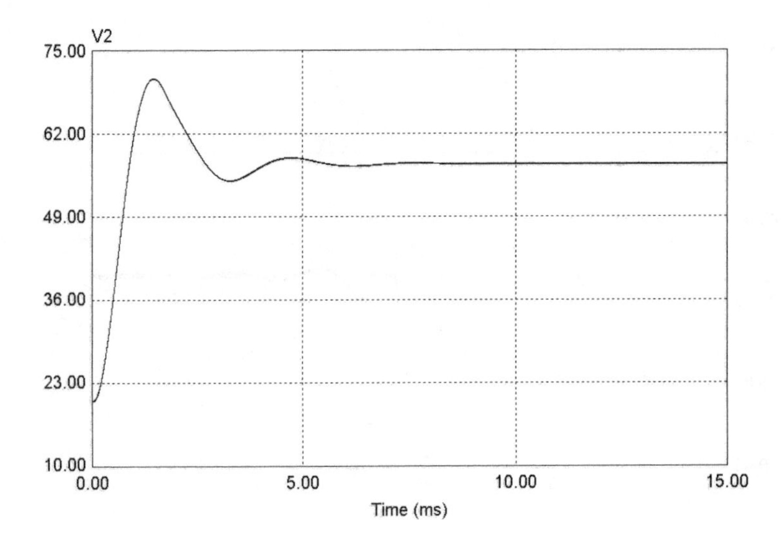

FIGURE 8.17
Super-lift Luo-converter unit-step responses.

The unit-step function response (transient process) has oscillation progress with damping factor σ and frequency ω. The simulation result is shown in Figure 8.17.

The impulse interference response is

$$\Delta v_2(t) = 0.932Ue^{\overline{0.00155}^{\,t}} \sin 1398t$$

where U is the interference signal. The impulse response (interference recovery process) has oscillation progress with damping factor σ and frequency ω, and is shown in Figure 8.18.

In order to verify the analysis, calculation, and simulation results; we constructed a test rig with same conditions. The corresponding test results are shown in Figures 8.19 and 8.20.

8.7.3 A Boost Converter in CCM (No Power Losses)

A boost converter shown in Figure 8.21 has the components values: V_1 = 40 *V*, *L* = 250 μH, *C* = 60 μF, *R* = 10 Ω, the switching frequency f = 20 kHz (*T* = 1/f = 50 μs) and conduction duty cycle k = 0.6. This converter is stable and works in CCM.

Therefore, we have got the voltage transfer gain $M = 1/(1 - k) = 2.5$, i.e. $V_2 = V_C = V_1 / (1 - k) = 100$ V. $I_2 = 10$ A, and $I_1 = I_L = 25$ A. The parameter *EF* and others are listed below:

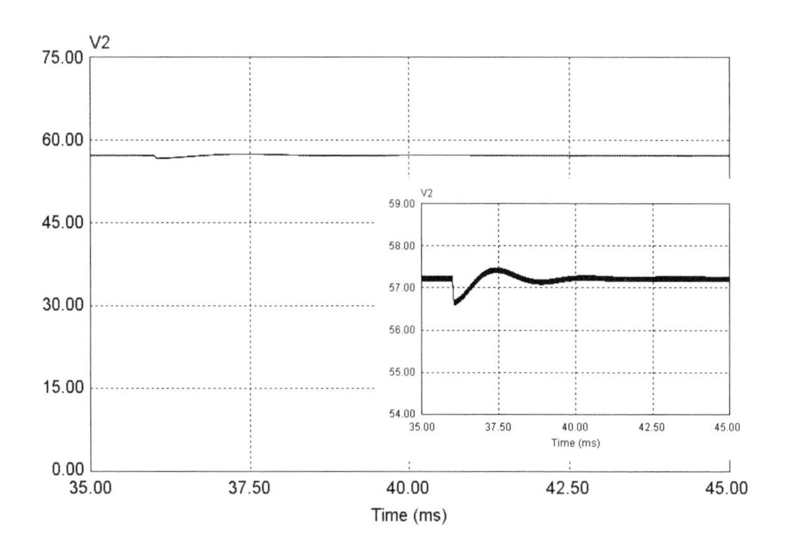

FIGURE 8.18
Super-lift Luo-converter impulse responses.

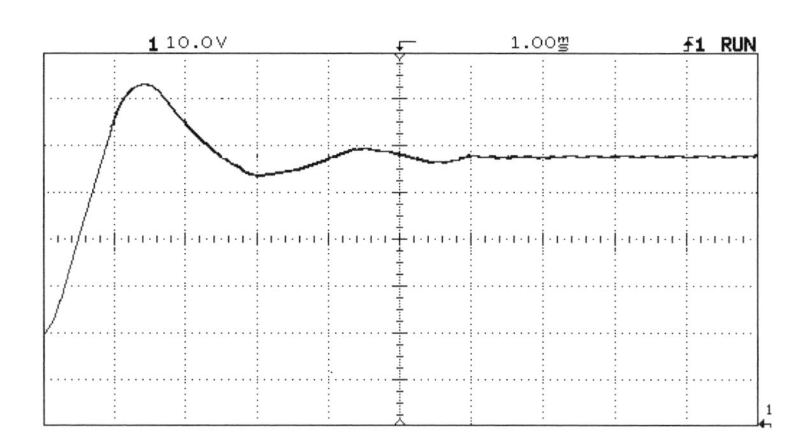

FIGURE 8.19
Unit-step function responses of super-lift Luo-converter (experiment).

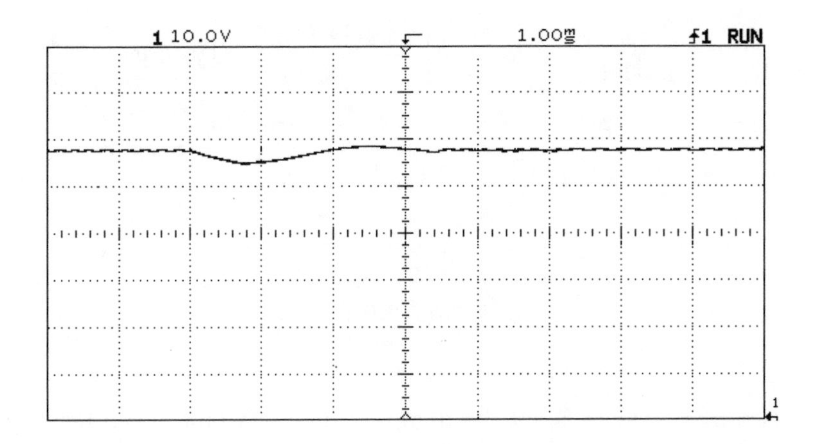

FIGURE 8.20
Impulse responses of super-lift Luo-converter (experiment).

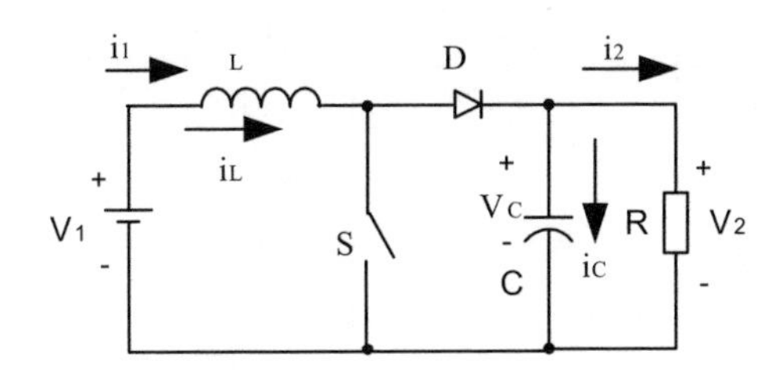

FIGURE 8.21
Boost converter.

$$PE = V_1 I_1 T = 40 * 25 * 50\mu = 50\,\text{mJ},\, W_L = \frac{1}{2} L I_L^2 = 0.5 * 250\mu * 25^2 = 78.125\,\text{mJ},$$

$$W_C = \frac{1}{2} C V_C^2 = 0.5 * 60\mu * 100^2 = 300\,\text{mJ},\, SE = W_L + W_C = 78.125 + 300 = 378.125\,\text{mJ},$$

$$EF = \frac{SE}{PE} = \frac{378.125}{50} = 7.5625 \qquad CIR = \frac{W_C}{W_L} = \frac{300}{78.125} = 3.84$$

Since no power losses in the converter, EL = 0 and $\eta = 1$:

$$\tau = \frac{2T \times EF}{1 + CIR}\left(1 + CIR\frac{1-\eta}{\eta}\right) = \frac{100\mu * 7.5625}{1 + 3.84} = 156.25\,\mu s$$

$$\tau_d = \frac{2T \times EF}{1 + CIR}\,\frac{CIR}{\eta + CIR(1-\eta)} = \frac{100 * 7.5625 * 3.84}{4.84} = 600\,\mu s$$

$$\xi = \frac{\tau_d}{\tau} = \frac{CIR}{\eta\left(1 + CIR\frac{1-\eta}{\eta}\right)^2} = 3.84 > 0.25$$

By cybernetic theory, since the damping time constant τ_d is much larger than the time constant τ, the corresponding ratio ξ is 3.84 > 0.25. The output voltage has heavy oscillation with high overshot. The transfer function of this converter has two poles ($-s_1$ and $-s_2$) that are located in the left-hand half plane (LHHP).

$$G(s) = \frac{M}{1 + s\tau + s^2\tau\tau_d} = \frac{M/\tau\tau_d}{(s + s_1)(s + s_2)} \tag{8.34}$$

where

$$s_1 = \sigma + j\omega \quad \text{and} \quad s_2 = \sigma - j\omega$$

with

$$\sigma = \frac{1}{2\tau_d} = \frac{1}{1.2ms} = 833 \text{ Hz}$$

$$\omega = \frac{\sqrt{4\tau\tau_d - \tau^2}}{2\tau\tau_d} = \frac{\sqrt{375000 - 24414}}{187500} = \frac{592.1}{187500\mu} = 3158 \text{ rad/s}$$

The unit-step function response is

$$v_2(t) = 100\left[1 - e^{\frac{t}{0.0012}}\left(\cos 3158t - 0.264 \sin 3158t\right)\right]V$$

The unit-step function response (transient process) has oscillation progress with damping factor σ and frequency ω. The simulation result is shown in Figure 8.22.

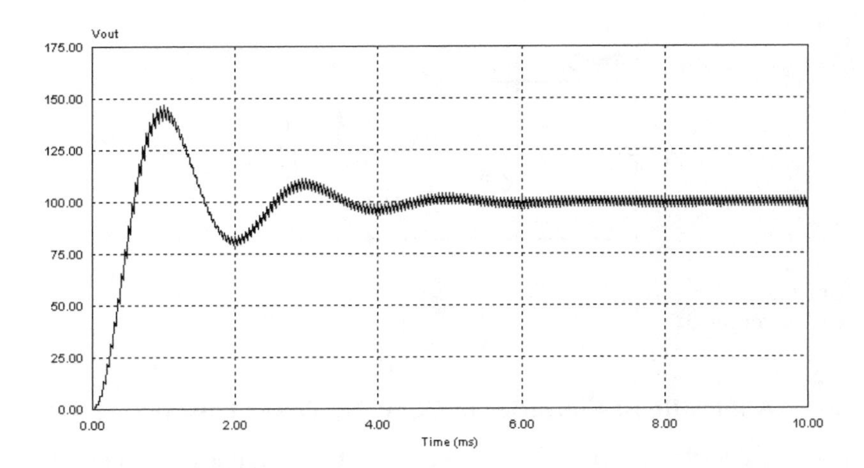

FIGURE 8.22
Boost converter unit-step responses.

The impulse interference response is

$$\Delta v_2(t) = 0.528Ue^{\frac{t}{0.0012}} \sin 3158t$$

where U is the interference signal. The impulse response (interference recovery process) has oscillation progress with damping factor σ and frequency ω, and is shown in Figure 8.23.

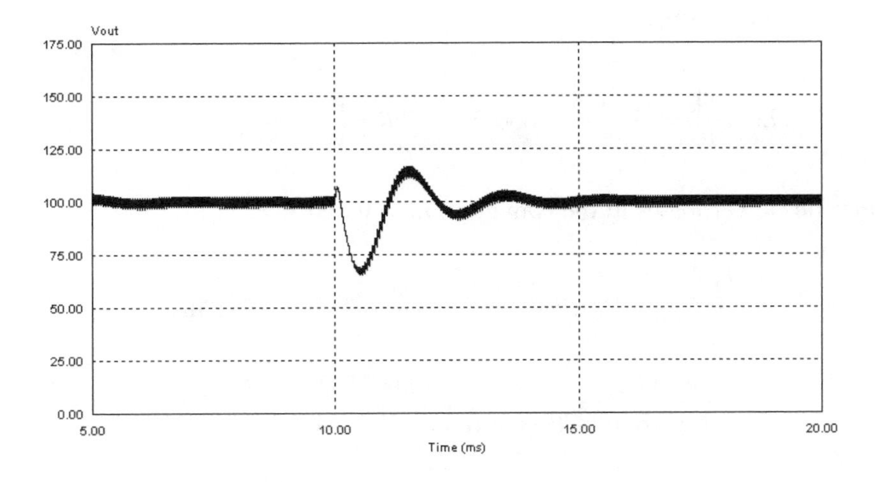

FIGURE 8.23
Boost converter impulse responses.

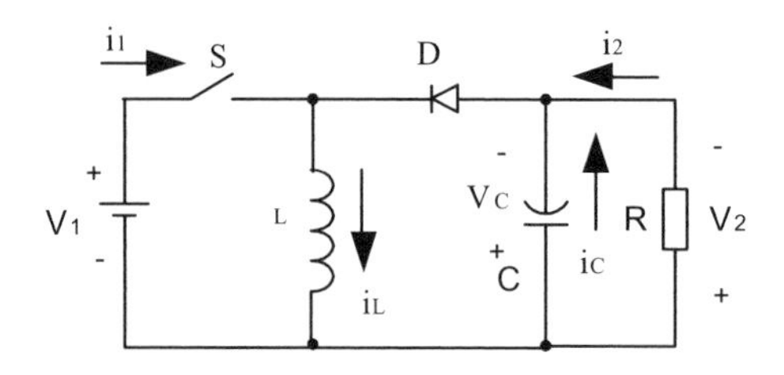

FIGURE 8.24
Buck-boost converter.

8.7.4 A Buck-Boost Converter in CCM (No Power Losses)

A Boost converter shown in Figure 8.24 has the components values: V_1 = 40 *V*, *L* = 250 μH, C = 60 μF, R = 10 Ω, the switching frequency f = 20 kHz ($T = 1/f$ = 50 μs) and conduction duty cycle k = 0.6. This converter is stable and works in CCM.

Therefore, we have got the voltage transfer gain $M = k/(1 - k) = 1.5$, i.e. $V_2 = V_C = kV_1 / (1 - k) = 60$ *V*. $I_2 = 6$ *A*, $I_1 = 9$ *A* and $I_L = 15$ *A*, The parameter *EF* and others are listed below:

$$PE = V_1 I_1 T = 40 * 9 * 50\mu = 18\,\text{mJ}, W_L = \frac{1}{2} L I_L^2 = 0.5 * 250\mu * 15^2 = 28.125\,\text{mJ},$$

$$W_C = \frac{1}{2} C V_C^2 = 0.5 * 60\mu * 60^2 = 108\,\text{mJ}, SE = W_L + W_C = 28.125 + 108 =$$

$$136.125\,\text{mJ},$$

$$EF = \frac{SE}{PE} = \frac{136.125}{18} = 7.5625 \qquad CIR = \frac{W_C}{W_L} = \frac{108}{28.125} = 3.84$$

Since no power losses in the converter, $EL = 0$ and $\eta = 1$:

$$\tau = \frac{2T \times EF}{1 + CIR}\left(1 + CIR\frac{1-\eta}{\eta}\right) = \frac{100\mu * 7.5625}{1 + 3.84} = 156.25\,\mu s$$

$$\tau_d = \frac{2T \times EF}{1 + CIR}\frac{CIR}{\eta + CIR(1-\eta)} = \frac{100 * 7.5625 * 3.84}{4.84} = 600\,\mu s$$

$$\xi = \frac{\tau_d}{\tau} = \frac{CIR}{\eta\left(1 + CIR\frac{1-\eta}{\eta}\right)^2} = 3.84 > 0.25$$

By cybernetic theory, since the damping time constant τ_d is much larger than the time constant τ, the corresponding ratio ξ is 3.84 > 0.25. The output voltage has heavy oscillation with high overshot. The transfer function of this converter has two poles ($-s_1$ and $-s_2$) that are located in the left-hand half plane (LHHP).

$$G(s) = \frac{M}{1 + s\tau + s^2\tau\tau_d} = \frac{M/\tau\tau_d}{(s + s_1)(s + s_2)} \tag{8.35}$$

where

$$s_1 = \sigma + j\omega \quad \text{and} \quad s_2 = \sigma - j\omega$$

with

$$\sigma = \frac{1}{2\tau_d} = \frac{1}{1.2ms} = 833 \text{ Hz}$$

$$\omega = \frac{\sqrt{4\tau\tau_d - \tau^2}}{2\tau\tau_d} = \frac{\sqrt{375000 - 24414}}{187500} = \frac{592.1}{187500\mu} = 3158 \text{ rad/s}$$

The unit-step function response is

$$v_2(t) = 60\left[1 - e^{\frac{t}{0.0012}}\left(\cos 3158t - 0.264\sin 3158t\right)\right] V$$

The unit-step function response (transient process) has oscillation progress with damping factor σ and frequency ω. The simulation result is shown in Figure 8.25.

The impulse interference response is

$$\Delta v_2(t) = 0.528Ue^{\frac{t}{0.0012}} \sin 3158t$$

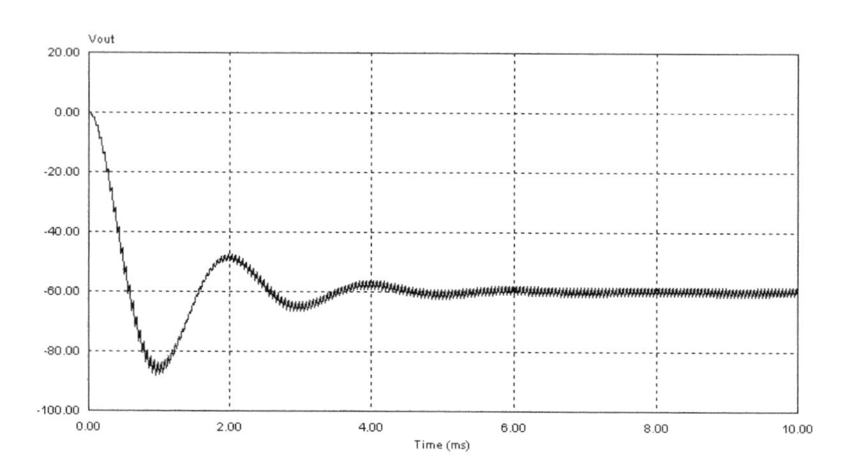

FIGURE 8.25
Buck-boost converter unit-step responses.

where U is the interference signal. The impulse response (interference recovery process) has oscillation progress with damping factor σ and frequency ω, and is shown in Figure 8.26.

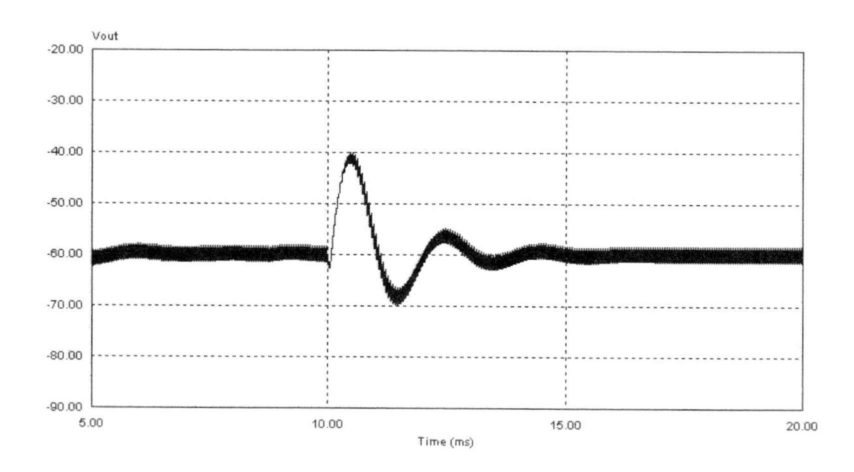

FIGURE 8.26
Buck-boost converter impulse responses.

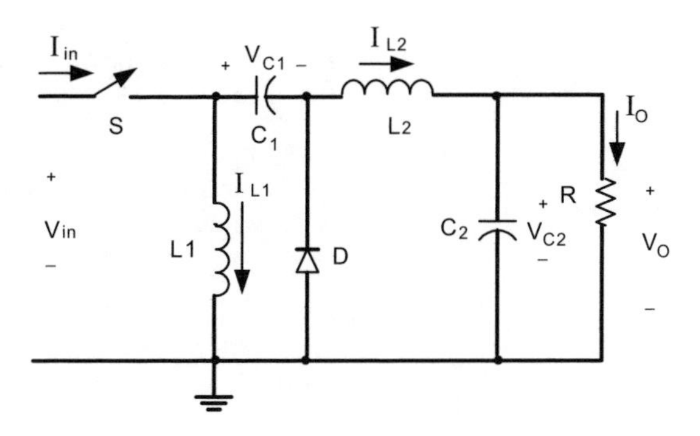

FIGURE 8.27
Positive-output Luo-converter.

8.7.5 Positive Output Luo-Converter in CCM (No Power Losses)

Figure 8.27 shows a positive output Luo-converter with the conduction duty is k. The components' values are V_1 = 20 V, f = 50 kHz (T = 20 μs), $L_1 = L_2$ = 1 mH, k = 0.5, $C_1 = C_2$ = 20 μF and R = 10 Ω. This converter is stable and works in *CCM*.

Therefore, we have got the voltage transfer gain $M = k/(1-k) = 1$, i.e., the output voltage $V_2 = V_{C2} = kV_1/(1-k) = 40\ V$. $V_{C1} = V_1 = 40\ V$, $I_1 = 4\ A$, $I_2 = 4\ A$ and $I_{L1} = I_{L2} = 4\ A$. The parameter *EF* and others are listed below:

$$PE = V_1 I_1 T = 40 * 4 * 20\mu = 3.2\,\text{mJ}, W_{L1} = \frac{1}{2} L_1 I_{L1}^2 = 0.5 * 1m * 4^2 = 8\,\text{mJ},$$

$$W_{L2} = \frac{1}{2} L_2 I_{L2}^2 = 0.5 * 1m * 4^2 = 8\,\text{mJ}, W_{C2} = \frac{1}{2} C_1 V_{C1}^2 = 0.5 * 20\mu * 40^2 = 16\,\text{mJ},$$

$$W_{C2} = \frac{1}{2} C_2 V_{C2}^2 = 0.5 * 20\mu * 40^2 = 16\,\text{mJ}, SE = W_{L1} + W_{L2} + W_{C1} + W_{C2} = 16 + 32 = 48\,\text{mJ},$$

$$EF = \frac{SE}{PE} = \frac{48}{3.2} = 15 \qquad CIR = \frac{W_{C1} + W_{C2}}{W_{L1} + W_{L2}} = \frac{32}{16} = 2$$

Since no power losses in the converter, EL = 0 and η = 1:

$$\tau = \frac{2T \times EF}{1+CIR}\left(1+CIR\frac{1-\eta}{\eta}\right) = \frac{40\mu * 15}{3} = 200\,\mu s$$

$$\tau_d = \frac{2T \times EF}{1+CIR}\frac{CIR}{\eta + CIR(1-\eta)} = \frac{40\mu * 15 * 2}{3} = 400\,\mu s$$

$$\xi = \frac{\tau_d}{\tau} = \frac{CIR}{\eta\left(1+CIR\frac{1-\eta}{\eta}\right)^2} = 2 > 0.25$$

By cybernetic theory, since the damping time constant τ_d is much larger than the time constant τ, the corresponding ratio ξ is 2 > 0.25. The output voltage has no oscillation and overshot. The transfer function of this converter has two real poles ($-s_1$ and $-s_2$) that are located in the left-hand half plane (LHHP).

$$G(s) = \frac{M}{1+s\tau+s^2\tau\tau_d} = \frac{M/\tau\tau_d}{(s+s_1)(s+s_2)} \tag{8.36}$$

where

$$s_1 = \sigma + j\omega \quad \text{and} \quad s_2 = \sigma - j\omega$$

with

$$\sigma = \frac{1}{2\tau_d} = \frac{1}{0.8ms} = 1250 \text{ Hz}$$

$$\omega = \frac{\sqrt{4\tau\tau_d - \tau^2}}{2\tau\tau_d} = \frac{\sqrt{320000-40000}}{160000} = \frac{529.2}{160000\mu} = 3307 \text{ rad/s}$$

The unit-step function response is

$$v_2(t) = 40\left[1 - e^{\frac{t}{0.0008}}\left(\cos 3307t - 0.378\sin 3307t\right)\right]V$$

The unit-step function response (transient process) has oscillation progress with damping factor σ and frequency ω. The simulation is shown in Figure 8.28.

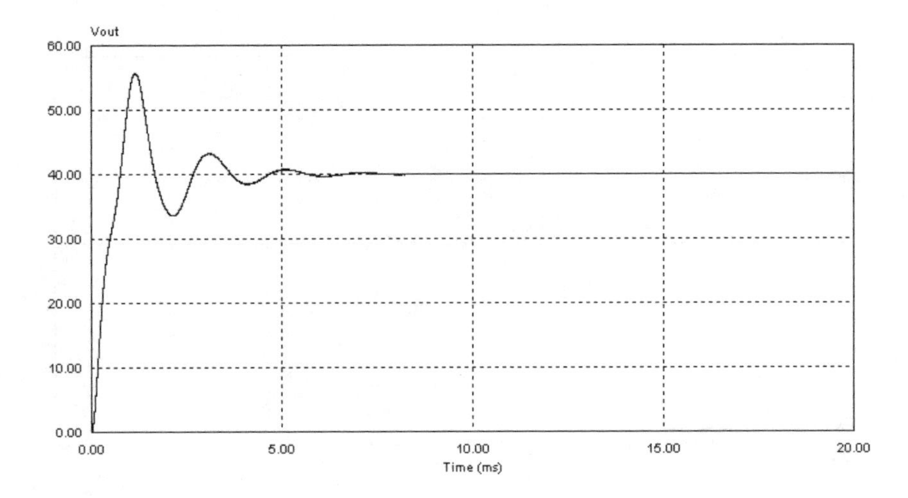

FIGURE 8.28
Positive-output Luo-converter unit-step responses.

The impulse interference response is

$$\Delta v_2\left(t\right) = 0.756Ue^{\frac{t}{0.0008}} \sin 3307t$$

where U is the interference signal. The impulse response (interference recovery process) has oscillation progress with damping factor σ and frequency ω, and is shown in Figure 8.29.

8.8 Small Signal Analysis

We analyzed the characteristics of power DC/DC converters in large signal operation in the previous section. We will analyze the characteristics of power DC/DC converters in small signal operation in this section. It will verify that the transfer function (8.28) is generally correct for both large and small signal analysis, and describe the native characteristics of a power DC/DC converter.

If the conduction duty cycle k changes from k_1 to k_2 ($\Delta k = k_2 - k_1$) in a small increment to the new value ($k_2 = k_1 + \Delta k$), the pumping energy *PE* changes correspondingly in an increment to the new value (*PE* + Δ*PE*). Analogously, the inductor currents and capacitor voltages have to change correspondingly, and the stored energy *SE* changes to (*SE* + Δ*SE*).

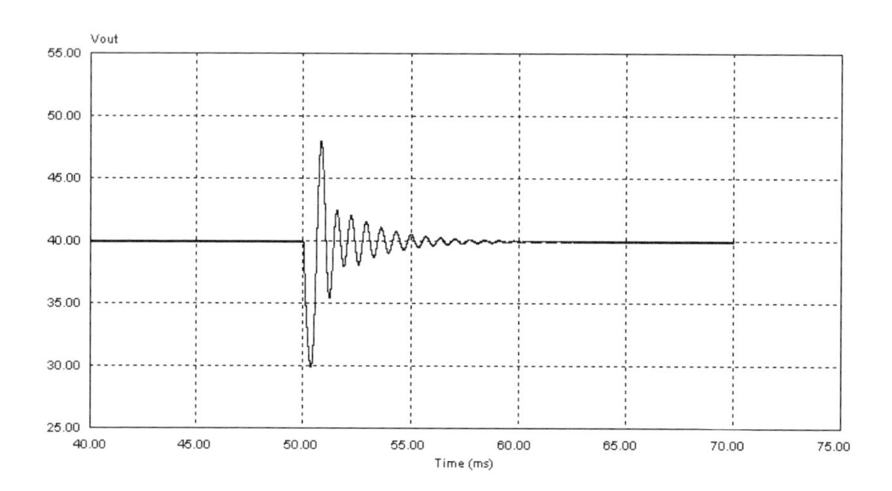

FIGURE 8.29
Positive-output Luo-converter impulse responses.

$$\Delta PE = \int_0^T V_1 i_1(t)\Big|_{k=k_2} dt - \int_0^T V_1 i_1(t)\Big|_{k=k_1} dt = V_1\left(I_{1-k_2} - I_{1-k_1}\right)T = V_1 \Delta I_1 T \tag{8.37}$$

The stored energy in an inductor is

$$\Delta W_L = \frac{1}{2} L\left(I_{L-k_2}^2 - I_{L-k_1}^2\right) \tag{8.38}$$

The stored energy across a capacitor is

$$\Delta W_C = \frac{1}{2} C\left(V_{C-k_2}^2 - V_{C-k_1}^2\right) \tag{8.39}$$

Therefore, if there are n_L inductors and n_C capacitors the total stored energy in a DC/DC converter is

$$\Delta SE = \sum_{j=1}^{n_L} \Delta W_{Lj} + \sum_{j=1}^{n_C} \Delta W_{Cj} \tag{8.40}$$

We define the energy factor EF in small signal operation as

$$EF = \frac{\Delta SE}{\Delta PE} = \frac{\sum_{j=1}^{m} \Delta W_{Lj} + \sum_{j=1}^{n} \Delta W_{Cj}}{V_1 \Delta I_1 T} \tag{8.41}$$

Correspondingly, the capacitor/inductor stored energy ratio (CIR) is

$$CIR = \frac{\sum_{j=1}^{n_C} \Delta W_{Cj}}{\sum_{j=1}^{n_L} \Delta W_{Lj}} \tag{8.42}$$

The energy losses increment (ΔEL) in a period T is defined as

$$\Delta EL = \Delta P_{loss} \times T \tag{8.43}$$

So that the efficiency η is.

$$\eta = \frac{\Delta PE - \Delta EL}{\Delta PE} \tag{8.44}$$

The time constant τ, damping time constant τ_d and time constant ratio ξ are not changed, they are still defined in the same forms:

$$\tau = \frac{2T \times EF}{1 + CIR}\left(1 + CIR\frac{1-\eta}{\eta}\right) \tag{8.45}$$

$$\tau_d = \frac{2T \times EF}{1 + CIR} \frac{CIR}{\eta + CIR(1-\eta)} \tag{8.46}$$

$$\xi = \frac{\tau_d}{\tau} = \frac{CIR}{\eta\left(1 + CIR\frac{1-\eta}{\eta}\right)^2} \tag{8.47}$$

The transfer function is not changed

$$G(s) = \frac{M}{1 + s\tau + s^2\tau\tau_d} = \frac{M}{1 + s\tau + \xi s^2\tau^2} \tag{8.48}$$

In order to verify this theory and offer examples to readers, we prepare two converters: a buck converter and super-lift Luo-converter below to demonstrate the characteristics of power DC/DC converters and applications of the theory.

8.8.1 A Buck Converter in CCM without Energy Losses ($r_L = 0$)

A buck converter shown in Figure 8.3 has the components values: $V_1 = 40$ V, $L = 250$ μH with resistance $r_L = 0$ Ω, $C = 60$ mF, $R = 10$ Ω, the switching frequency f = 20 kHz ($T = 1/f = 50$ μs) and conduction duty cycle k changing from 0.4 to 0.5. This converter is stable and works in *CCM*.

Therefore, we have got the voltage transfer gain $M = 0.5$, i.e. $V_2 = V_C = MV_1 = 0.5 \times 40 = 20$ V. $I_L = I_2 = 2$ A, $P_{loss} = 0$ W and $I_1 = 1$ A. The increments are $\Delta V_2 = 4$ V, $\Delta I_2 = \Delta I_L = 0.4$ A, $\Delta I_1 = 0.34$ A. The parameter *EF* and others are listed below:

$$\Delta PE = V_1 \Delta_1 T = 40 \times 0.36 \times 50\mu = 0.72 \text{ mJ},$$

$$\Delta W_L = \frac{1}{2} L\left(I^2_{L-0.5} - I^2_{L-0.4}\right) = 0.5 \times 250\mu \times \left(2^2 - 1.6^2\right) = 0.18 \text{ mJ},$$

$$\Delta W_C = \frac{1}{2} C\left(V^2_{C-0.5} - V^2_{C-0.4}\right) = 4.32 \text{ mJ}, \ \Delta \text{SE} = \Delta \text{W}_\text{L} + \Delta W_C = 4.5 \text{ mJ},$$

$$\Delta EL = 0 \text{ mJ}, \quad \eta = \frac{\Delta PE - \Delta EL}{\Delta PE} = 1$$

$$EF = \frac{\Delta SE}{\Delta PE} = \frac{4.5}{0.72} = 6.25, \quad CIR = \frac{\Delta W_C}{\Delta W_L} = \frac{4.32}{0.18} = 24$$

$$\tau = \frac{2T \times EF}{1 + CIR}\left(1 + CIR \frac{1-\eta}{\eta}\right) = 25\,\mu s, \ \tau_d = \frac{2T \times EF}{1 + CIR} \frac{CIR}{\eta + CIR(1-\eta)} = 600\,\mu s$$

From the above calculation and analysis, we found that the time constants are unchanged. Therefore, the transfer function for small signal operation should not be changed and is still (8.29). Correspondingly, the unit-step response is

$$v_2(t) = 16 + 4\left[1 - e^{\frac{t}{0.0012}}\left(\cos 8122t - 0.1026 \sin 8122t\right)\right] V$$

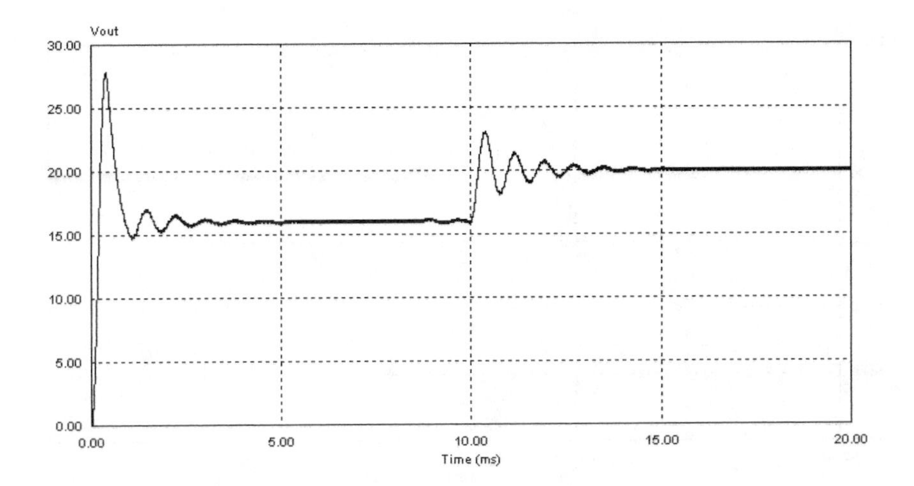

FIGURE 8.30
Unit-step responses of buck converter without power loss (simulation).

The unit-step function response for large signal ($k = 0$ to 0.4) and small signal ($k = 0.4$ to 0.5) operation is shown in Figure 8.30 for comparison with each other.

The impulse response for small signal is described by,

$$\Delta v_2(t) = 0.20Ue^{\frac{t}{0.0012}} \sin 8122t$$

where U is the interference signal. The small-signal impulse response is shown in Figure 8.31.

In order to verify this analysis and compare the simulation results to experimental results, a test rig was constructed. The conduction duty cycle k changes from k = 0.4 to 0.5. The experimental results for unit-step response (k = 0.4 to 0.5) and impulse interference response (k = 0.5 to 0.4) are shown in Figures 8.32 and 8.33. We can see that both simulation and experimental results are identical to each other.

8.8.2 Buck-Converter with Small Energy Losses (r_L = 1.5 Ω)

The buck converter shown in Figure 8.3 has the components values: V_1 = 40 *V*, L = 250 μH with resistance r_L = 1.5 Ω, C = 60 μF, R = 10 Ω, the switching frequency f = 20 kHz ($T = 1/f$ = 50 μs) and conduction duty cycle k changing from 0.4 to 0.5. This converter is stable and works in *CCM*.

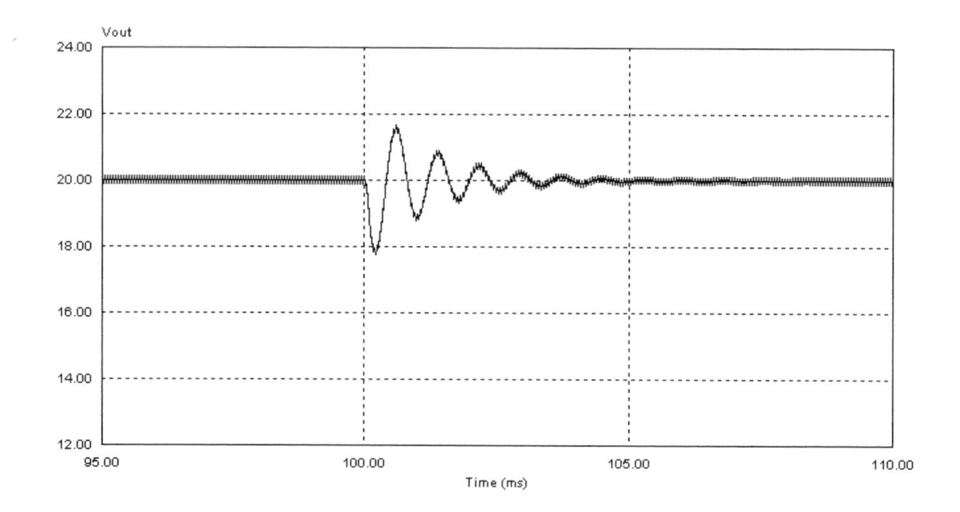

FIGURE 8.31
Impulse responses of buck converter without power loss (small signal).

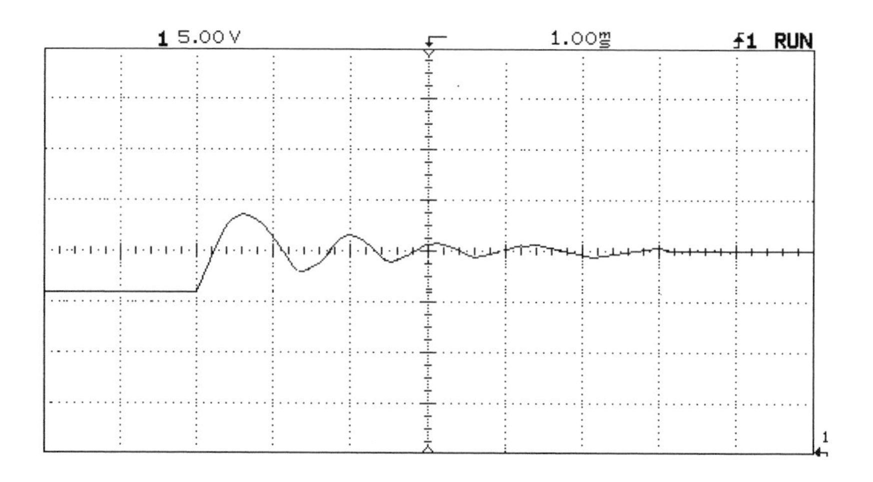

FIGURE 8.32
Unit-step responses of buck converter without power losses (experiment, small signal).

We have the voltage transfer gain $M = 0.435$, i.e. $V_2 = V_C = MV_1 = 0.435 \times 40 = 17.4$ V. $I_L = I_2 = 17.4$ A, $P_{loss} = I^2_L\, r_L = 1.74^2 \times 1.5 = 4.54$ W, $I_1 = 0.871$ A, and $\Delta V_2 = 3.4$ V, $\Delta I_2 = \Delta I_L = 0.34$ A, $\Delta P_{loss} = (I^2_{L\text{-}0.5} - I^2_{L\text{-}0.4})\, r_L = (1.74^2 - 1.4^2) \times 1.5 = 1.6$ W, $\Delta I_1 = 0.313$ A. The parameters are listed below:

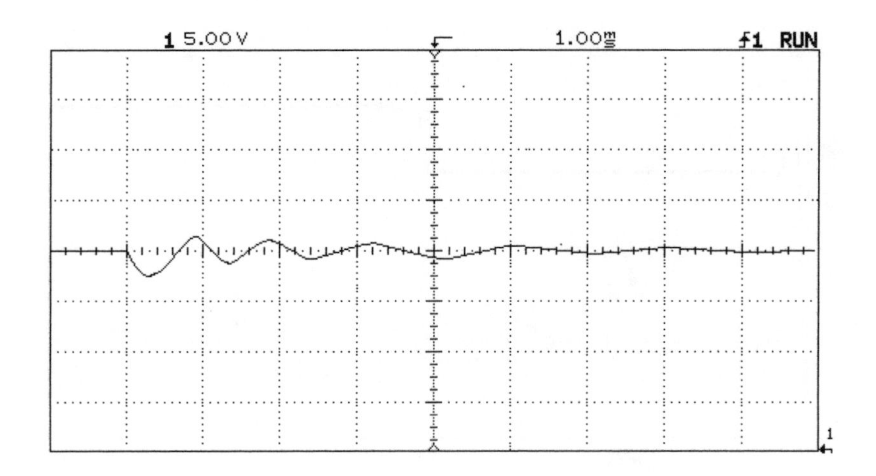

FIGURE 8.33
Small signal impulse responses of buck converter without power losses (experiment).

$$\Delta PE = V_1 \Delta I_1 T = 40 \times 0.313 \times 50\mu = 0.626 \text{ mJ}, \Delta W_L = \frac{1}{2} L\left(I_{L-0.5}^2 - I_{L-0.4}^2\right) = 0.136 \text{ mJ},$$

$$\Delta W_C = \frac{1}{2} C\left(V_{C-0.5}^2 - V_{C-0.5}^2\right) = 3.284 \text{ mJ}, \Delta SE = \Delta W_L + \Delta W_C = 3.42 \text{ mJ},$$

$$\Delta EL = \Delta P_{loss} * T = 1.6 * 50 = 0.08 \text{ mJ}, \eta = \frac{\Delta PE - \Delta EL}{\Delta PE} = \frac{0.546}{0.626} = 0.87$$

$$EF = \frac{\Delta SE}{\Delta PE} = \frac{3.42}{0.626} = 5.43, \quad CIR = \frac{\Delta W_C}{\Delta W_L} = \frac{3.284}{0.136} = 24$$

$$\tau = \frac{2T \times EF}{1 + CIR}\left(1 + CIR \frac{1-\eta}{\eta}\right) = 100\,\mu s, \tau_d = \frac{2T \times EF}{1 + CIR} \frac{CIR}{\eta + CIR(1-\eta)} = 130.6\,\mu s$$

From the above calculation and analysis, we found that the time constants are unchanged. Therefore, the transfer function for small signal operation should not be changed and is still (8.30). Correspondingly, the unit-step response is

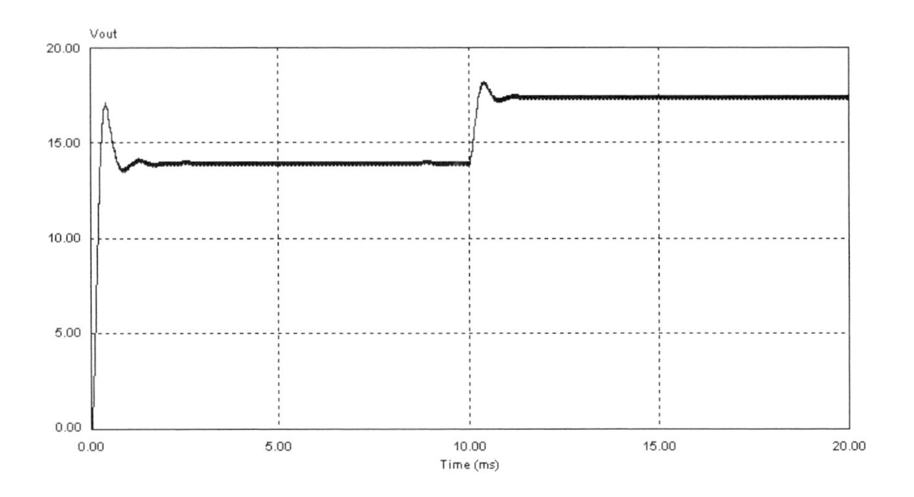

FIGURE 8.34
Unit-step responses (simulation) of buck converter with power loss (r_L = 1.5 Ω).

$$v_2(t) = 14 + 3.4\left[1 - e^{\frac{t}{0.000261}}\left(\cos 7888t - 0.486 \sin 7888t\right)\right] V$$

The unit-step function response for large (k = 0 to 0.4) and small signal (k = 0.4 to 0.5) operation is shown in Figure 8.34 for comparison with each other. The impulse response for small signal is described by

$$\Delta v_2(t) = 0.973Ue^{\frac{t}{0.000261}} \sin 7888t$$

where U is the interference signal. The small-signal impulse response is shown in Figure 8.35.

In order to verify this analysis and compare the simulation results to experimental results, a test rig was constructed. The conduction duty cycle k changes from k = 0.4 to 0.5. The experimental results for unit-step response (k = 0.4 to 0.5) and impulse interference response (k = 0.5 to 0.4) are shown in Figures 8.36 and 8.37.

We can see that both simulation and experimental results are identical to each other.

8.8.3 Super-Lift Luo-Converter with Energy Losses (r_L = 0.12 Ω)

Figure 8.16 shows a super-lift Luo-converter with the conduction duty k changing from 0.5 to 0.6. The components values are V_1 = 20 V, f = 50 kHz

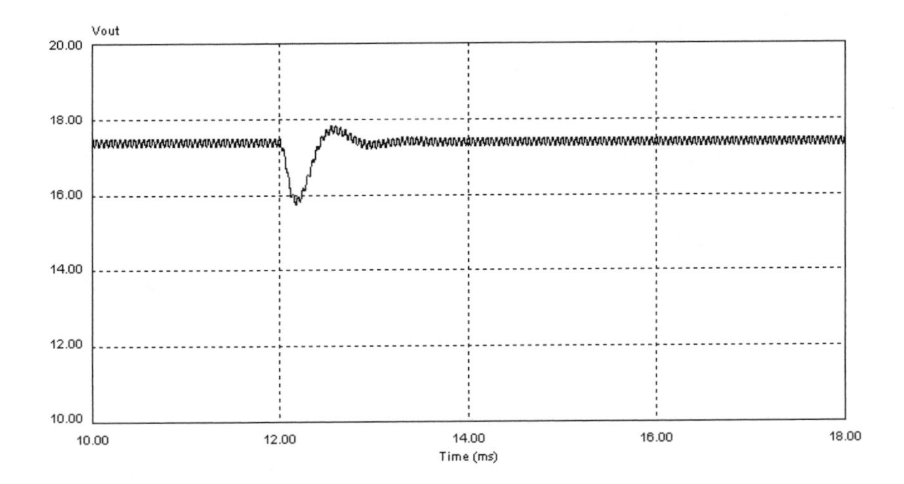

FIGURE 8.35
Impulse responses (simulation) of buck converter with power loss (r_L = 1.5 Ω, small signal).

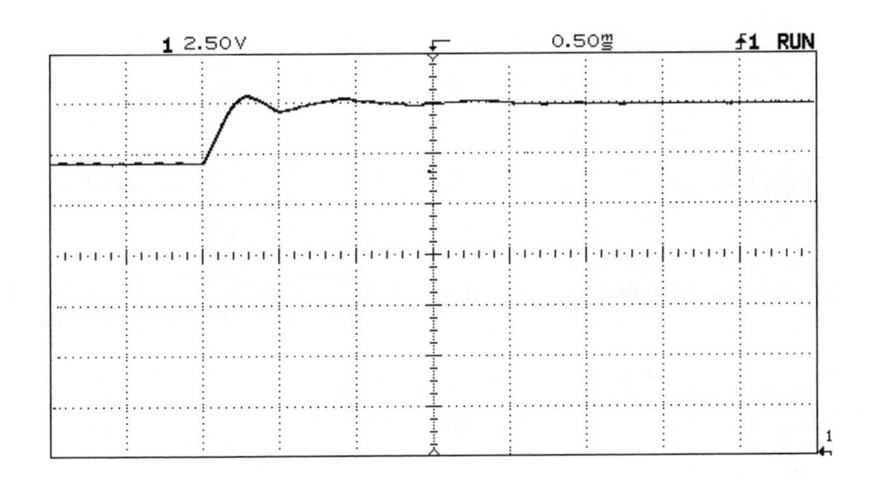

FIGURE 8.36
Unit-step function responses (experiment, small signal) of buck converter with power loss (r_L = 1.5 Ω).

(T = 20 μs), L = 100 μH with resistance r_L = 0.12 Ω, C_1 = 2500 μF, C_2 = 800 μF and R = 10 Ω. This converter is stable and works in CCM.

We then obtain $V_2 = 65.09\ V$, I_2 = 6.509 A, I_1 = 22 A, I_L = 14.91 A, $P_{loss} = I^2_L \times r_L = 14.91^2 \times 0.12 = 26.67\ W$, $V_{C1} = V_1 = 20\ V$, $V_{C2} = V_2 = 65.09\ V$, and $\Delta V_2 = 7.74\ V$, $\Delta I_2 = 0.784\ A$, $\Delta I_1 = 4.825\ A$, $\Delta I_L = 3.46\ A$, $\Delta P_{loss} = (I^2_{L\text{-}0.6} - I^2_{L\text{-}0.5})\ r_L = 7.4\ W$, $\Delta V_{C1} = \Delta V_1 = 0\ V$, $\Delta V_{C2} = \Delta V_2 = 7.84\ V$. The parameters are

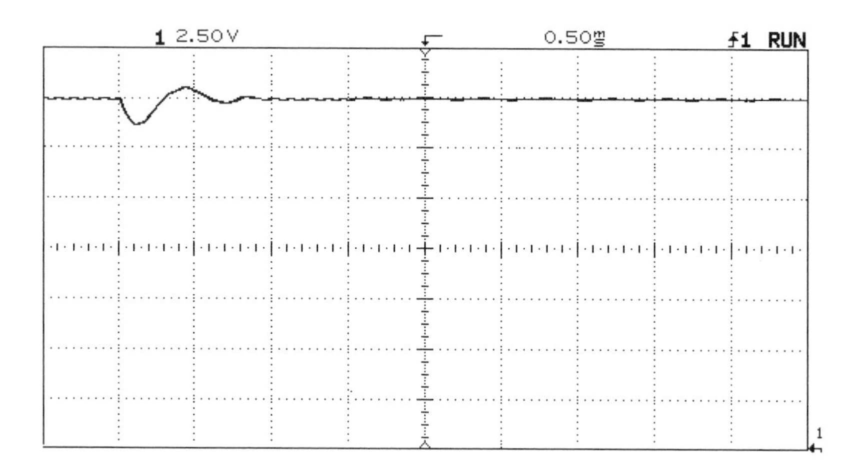

FIGURE 8.37
Impulse responses (experiment, small signal) of buck converter with power loss (r_L = 1.5 Ω).

$$\Delta PE = V_1 \Delta I_1 T = 1.93 \text{ mJ}, \quad \Delta W_L = 1.385 \text{ mJ},$$

$$\Delta W_{C1} = 0 \text{ mJ}, \quad \Delta W_{C2} = 383.68 \text{ mJ},$$

$$\Delta SE = \Delta W_L + \Delta W_{C1} + \Delta W_{C2} = 385.06 \text{ mJ}, \; EF = \frac{\Delta SE}{\Delta PE} = \frac{385.06}{1.9} = 203$$

$$\Delta EL = \Delta P_{loss} * T = 7.4 * 20 = 0.148 \text{ mJ}, \; \eta = \frac{\Delta PE - \Delta EL}{\Delta PE} = \frac{1.93 - 0.148}{1.93} = 0.923$$

$$CIR = \frac{\Delta W_C}{\Delta W_L} = \frac{383.68}{1.38} = 277$$

$$\tau = \frac{2T \times EF}{1 + CIR}\left(1 + CIR\frac{1-\eta}{\eta}\right) = 543 \text{ µs}$$

$$\tau_d = \frac{2T \times EF}{1 + CIR}\frac{CIR}{\eta + CIR(1-\eta)} = 768 \text{ µs}$$

From the above calculation and analysis we found that the time constants are unchanged. Therefore, the transfer function (8.33) for small signal operation should not be changed. Correspondingly, the unit-step response is

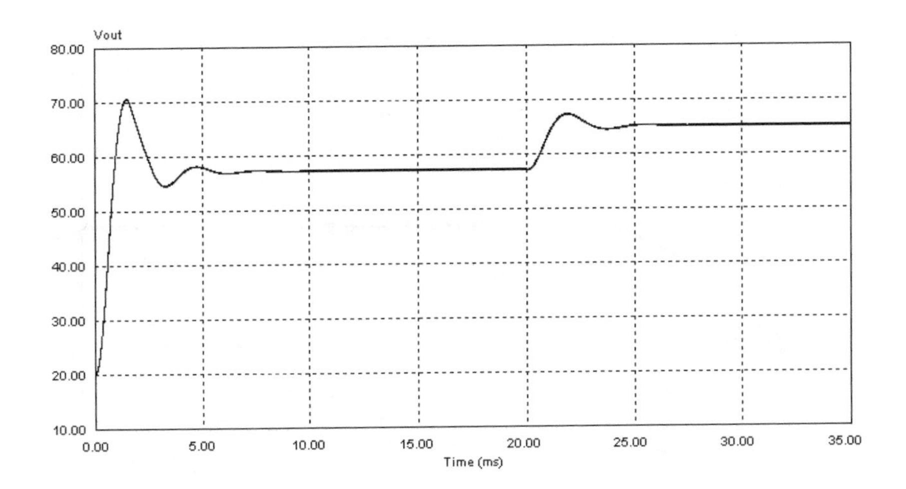

FIGURE 8.38
Unit-step responses (simulation) of super-lift Luo-converter (r_L = 0.12 Ω).

$$v_2(t) = 57.25 + 7.8\left[1 - e^{\frac{t}{1.55}}\left(\cos 1398t - 0.461 \sin 1398t\right)\right] V$$

The unit-step function response for large signal (k = 0 to 0.5) and small signal (k = 0.5 to 0.6) operations are shown in Figure 8.38. The impulse response for small signal is described by

$$\Delta v_2(t) = 0.923Ue^{\frac{t}{0.00155}} \sin 1398t$$

where U is the interference signal. The small signal impulse response is shown in Figure 8.39.

In order to verify this analysis and compare the simulation results to experimental results, a test rig was constructed. The components' values are V_1 = 20 V, f = 50 kHz (T = 20 μs), k = 0.5, L = 100 μH (with r_L = 0.12 Ω), C_1 = 750 μF, C_2 = 200 μF and R = 10 Ω.

The experimental results for unit-step (small signal: k = 0.5 to 0.6) response and impulse interference (small signal: k = 0.6 to 0.5) responses are shown in Figures 8.40 and 8.41. We can see that both simulation and experimental results are identical to each other.

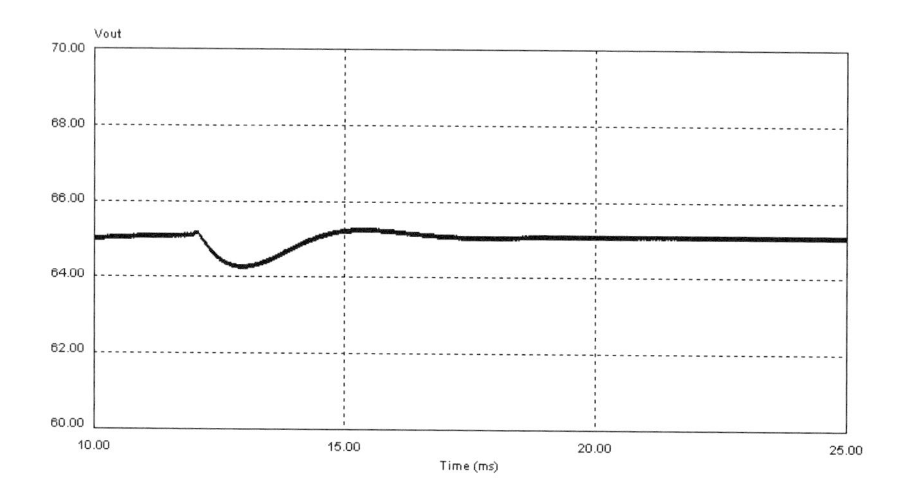

FIGURE 8.39
Impulse responses (simulation) of super-lift Luo-converter (r_L = 0.12 Ω, small signal).

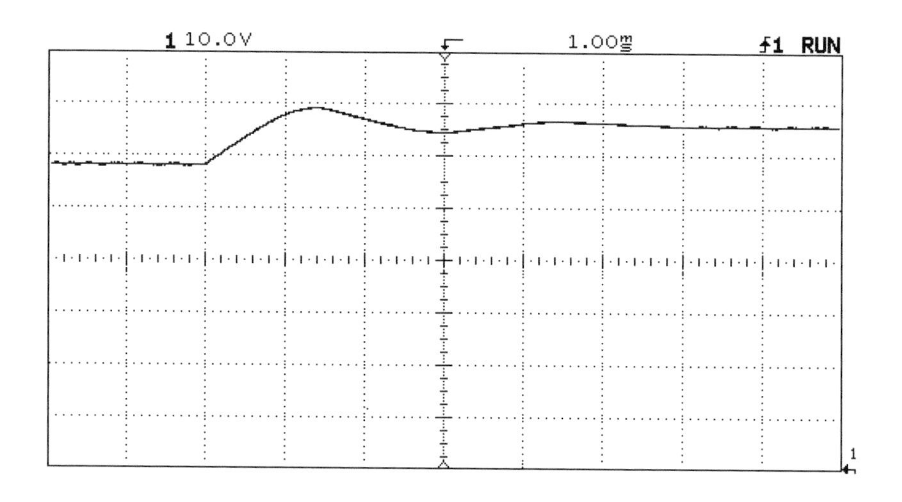

FIGURE 8.40
Unit-step function responses (experiment, small signal) of super-lift Luo-converter (r_L = 0.12 Ω).

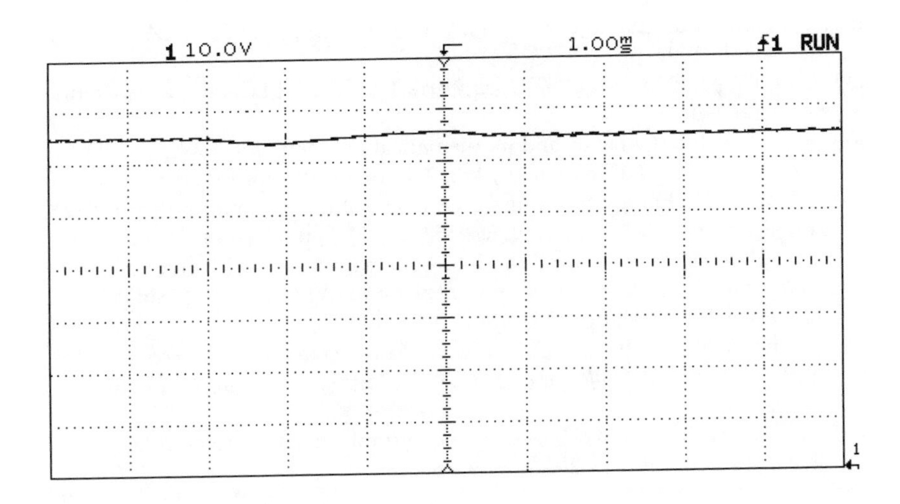

FIGURE 8.41
Impulse responses (experiment, small signal) of super-lift Luo-converter (r_L = 0.12 Ω).

Bibliography

Cheng K. W. E., Storage energy for classical switched mode power converters *IEE-Proceedings on EPA*, 150, 4, 2003, pp. 439–446.

Czarkowski, D. and Kazimierczuk M. K., Energy conservation approach to modelling PWM DC-DC converters *IEEE Trans. on Aerospace and Electronic Systems*, 29, 3, July 1993, pp. 1059–1063.

Dariusz Czarkowski, Pujara L. R., and Kazimierczuk M. K., Robust stability of state-feedback control of PWM DC-DC push-pull converter, *IEEE Trans. on IE*, 42, 1, February 1995, pp. 108–111.

Erickson R. W. and Maksimovic D., *Fundamentals of Power Electronics*, 2nd ed., Boulder, Colorado, Kluwer Academic Publishers, 2001.

Kaszimierczuk, M. K. and Cravens II, R, Closed-loop characteristics of voltage-mode controlled PWM boost converter with an intergal-lead controller, *Journal of Circuits, Systems and Computers*, 4, 4, December 1994, pp. 429–458.

Kazimierczuk, M. K. and R. Cravens II, Experimental results for the small-signal study of the PWM boost converter with an intergral-lead controller, *Journal of Circuits, Systems and Computers*, 5, 4, December 1995, pp. 747–755.

Kazimierczuk, M.K. and Cravens II, R., Open and closed-loop dc and small-signal characteristics of PWM buck-boost converter for CCM, *Journal of Circuits, Systems and Computers*, 5, 3, September 1995, pp. 261–3003.

Lee Y. S., A systemic and unified approach to modeling switches in switch-mode power supplies *IEEE-Trans. on IE*, 32, 1985, pp. 445–448.

Luo F. L., Double output Luo-converters: Advanced voltage lift technique, *IEE-EPA Proceedings* 147, 6, November 2000, pp. 469–485.

Luo F. L., Negative output Luo-converters: Voltage lift technique, *IEE-EPA Proceedings*, 146, 2, March 1999, pp. 208–224.

Luo F. L., Positive output Luo-converters: Voltage lift technique, *IEE-EPA Proceedings*, 146, 4, July 1999, pp. 415–432.
Luo F. L. and Ye H., *Advanced DC/DC converters*, CRC Press LLC, Boca Rotan, Florida, September 2003.
Luo F. L. and Ye, Energy factor and mathematical modeling for power DC/DC converters, *IEE-EPA Proceedings*, 152, 2, March 2005, pp. 191–198.
Luo F. L. and Ye H., Energy factor and mathematical modeling for power DC/DC converters, in *IEE-EPA Proceedings*, 152, 2, 2005, pp. 233–248.
Luo F. L. and Ye H., Mathematical modeling for power DC/DC converters, in *Proceedings of the IEEE International Conference POWERCON'2004*, Singapore, 21-24, November, 2004, pp. 323–328.
Luo F. L. and Ye H., Positive output super-lift Luo-converters, in *Proceedings of the IEEE International Conference PESC'2002*, Cairns, Australia, 23–27, June 2002, pp. 425–430.
Luo F., Ye H., and Rashid M. H., Multiple-quadrant Luo-converters, *IEE-EPA Proceedings* 148, 1, January 2002, pp. 9–18.
Luo F. L. and Ye H., Negative output super-lift Luo-converters, *Proceedings of the IEEE International Conference PESC'2003*, Acapulco, Mexico, 15–19, June 2003, pp. 1361–1366.
Luo F. L. and Ye H., Positive output super-lift converters, *IEEE-Trans. on PEL*, 18, 1, January 2003, pp. 105–113.
Luo F. L. and Ye H., Negative output super-lift converters, *IEEE-Trans. on PEL*, 18, 5, September 2003, pp. 1113–1121.
Luo F. L. and Ye H., Positive output cascade boost converters, *IEE-EPA Proceedings*, 151, 5, September 2004, pp. 590–606.
Middlebrook R. and Cúk, S., A general unified approach to modeling switching-converter power stages, *Journal of Electronics*, 42, 6, June 1977, pp. 521–550.
Sira-Ramirez, H., A geometric approach to pulse-width modulated control in non-linear dynamical systems, *IEEE Trans. on Auto-Cont*, 34, 2, February 1989, pp. 184–187.
Sira-Ramirez, H., Sliding motions in bilinear switched networks, *IEEE Trans. on CAS*, 34, 8, August 1987, pp. 919–933.
Sira-Ramirez, H. and Ilic, M., Exact linearization in switched mode DC to DC power converters, *International Journal of Control*, 50, 2, August 1989, pp. 511–524.
Sira-Ramirez, H., Ortega, R., Perez-Moreno, R., and Garcia-Esteban, M., A sliding mode controller-observer for DC-to-DC power converters: A passivity approach, *Proceedings of IEEE-DAC'95*, 4, 1995, pp. 3379–3384.
Sira-Ramirez, H. and Rios-Bolivar Miguel, Sliding mode control of DC-to-DC power converters via extended linearization, *IEEE Trans. on CAS*, 41, 10, October 1994, pp. 652–661.
Smedley, K. M. and Cuk, S., One-cycle control of switching converters, *IEEE Trans. on PEL*, 10, 6, November 1995, pp. 625–633
Wong R. C., Owen H. A. and Wilson T. G., An efficient algorithm for the time-domain simulation of regulated energy-storage DC-to-DC converters, *IEEE-Trans. on PEL*, 2, 1987, pp. 54–168.

Appendix A: A Second-Order Transfer Function

A typical second-order transfer function in the s-domain is shown below:

$$G(s) = \frac{M}{1 + s\tau + s^2\tau\tau_d} = \frac{M}{1 + s\tau + \xi s^2\tau^2} \tag{8A.1}$$

where M voltage transfer gain
τ The time constant
τ_d The damping time constant, $\tau_d = \xi\tau$
s The Laplace operator in s-domain

We now discuss various situations for the transfer function in detail.

A1 Very Small Damping Time Constant

If the damping time constant is very small (i.e. $\tau_d << \tau$, $\xi << 1$) and it can be ignored, the value of the damping time constant τ_d is omitted (i.e. $\tau_d = 0$, $\xi = 0$). The transfer function (19A.1) is downgraded to the first order as

$$G(s) = \frac{M}{1 + s\tau} \tag{8A.2}$$

The unit-step function response in the time-domain is

$$g(t) = M\left(1 - e^{-\frac{1}{\tau}}\right) \tag{8A.3}$$

The transient process (settling time) is nearly three times the time constant, 3τ to produce $g(t) = g(3\tau) = 0.95M$. The response in time-domain is shown in Figure 19A.1 with $\tau_d = 0$.

The impulse interference response is

$$\Delta g(t) = U \cdot e^{-\frac{t}{\tau}} \tag{8A.4}$$

where U is the interference signal. The interference recovering progress is nearly three times of the time constant, 3τ, and is shown in Figure 8A.2 with $\tau_d = 0$.

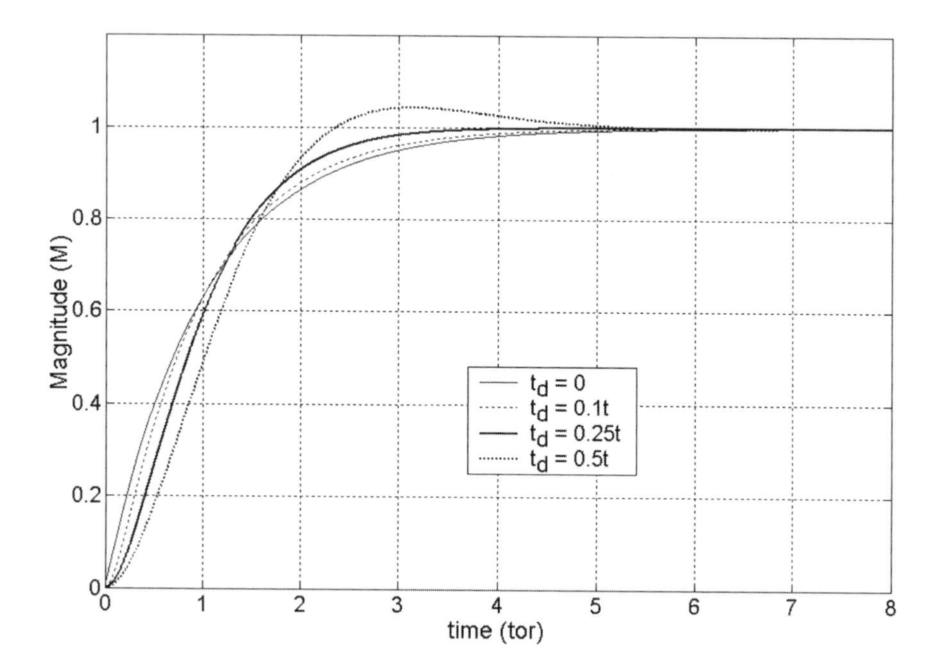

FIGURE 8A.1
Unit-step function responses (τ_d = 0, 0.1τ, 0.25τ and 0.5τ).

A2 Small Damping Time Constant

If the damping time constant is small (i.e. $\tau_d < \tau/4$, $\xi < 0.25$) and it cannot be ignored, the value of the damping time constant τ_d is not omitted. The transfer function (1) is retained as the second-order function with two real poles $-\sigma_1$ and $-\sigma_2$ as

$$G(s) = \frac{M}{1 + s\tau + s^2\tau\tau_d} = \frac{M/\tau\tau_d}{(s+\sigma_1)(s+\sigma_2)} \tag{8A.5}$$

where

$$\sigma_1 = \frac{\tau + \sqrt{\tau^2 - 4\tau\tau_d}}{2\tau\tau_d} \quad \text{and} \quad \sigma_2 = \frac{\tau - \sqrt{\tau^2 - 4\tau\tau_d}}{2\tau\tau_d}$$

There are two real poles in the transfer function, assuming $\sigma_1 > \sigma_2$. The unit-step function response in the time-domain is

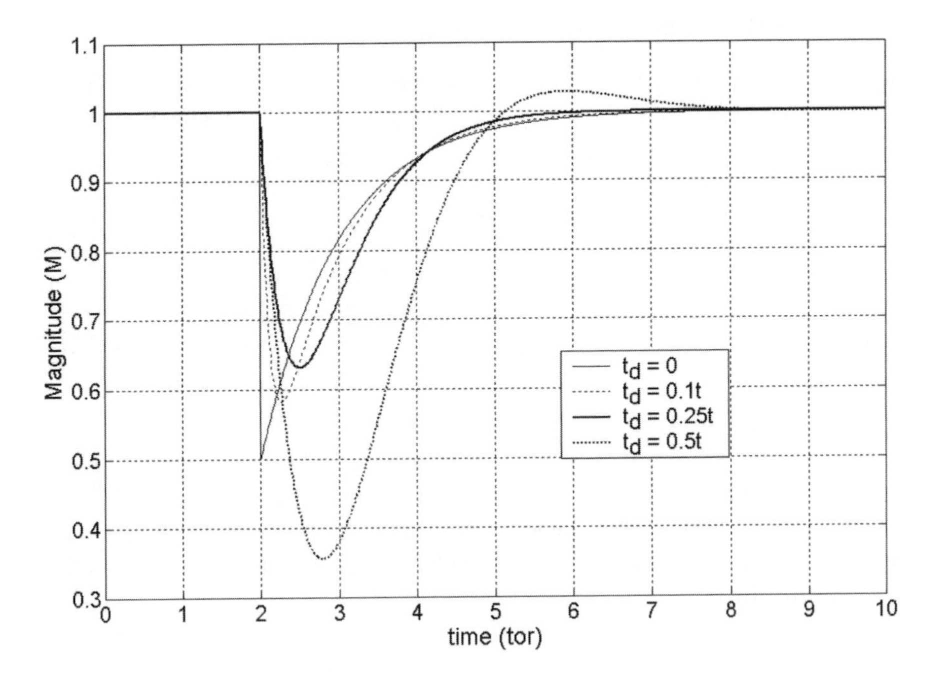

FIGURE 8A.2
Impulse responses(τ_d = 0, 0.1τ, 0.25τ and 0.5τ).

$$g(t) = M(1 + K_1 e^{-\sigma_1 t} + K_2 e^{-\sigma_2 t}) \tag{8A.6}$$

where

$$K_1 = -\frac{1}{2} + \frac{\tau}{2\sqrt{\tau^2 - 4\tau\tau_d}} \qquad K_1 = -\frac{1}{2} - \frac{\tau}{2\sqrt{\tau^2 - 4\tau\tau_d}}$$

The transient process is nearly three times the time value $1/\sigma_1$, $3/\sigma_1 < 3\tau$. The response process is quick without oscillation. The corresponding waveform in time-domain is shown in Figure 8A.1 with $\tau_d = 0.1\tau$.

The impulse interference response is

$$\Delta g(t) = \frac{U}{\sqrt{1 - 4\tau_d/\tau}}\left(e^{-\sigma_2 t} - e^{-\sigma_1 t}\right) \tag{8A.7}$$

where U is the interference signal. The transient process is nearly three times the time value $1/\sigma_1$, $3/\sigma_1 < 3\tau$. The response waveform in time-domain is shown in Figure 8A.2 with $\tau_d = 0.1\tau$.

A3 Critical Damping Time Constant

If the damping time constant is equal to the critical value (i.e. $\tau_d = \tau/4$), the transfer function (8A.1) is retained as the second-order function with two equaled real poles $\sigma_1 = \sigma_2 = \sigma$ as

$$G(s) = \frac{M}{1 + s\tau + s^2\tau\tau_d} = \frac{M/\tau\tau_d}{(s+\sigma)^2} \tag{8A.8}$$

where

$$\sigma = \frac{1}{2\tau_d} = \frac{2}{\tau}$$

There are two folded real poles in the transfer function. This expression describes the characteristics of the DC/DC converter. The unit-step function response in the time-domain is

$$g(t) = M\left[1 - \left(1 + \frac{2t}{\tau}\right)e^{\frac{2t}{\tau}}\right] \tag{8A.9}$$

The transient process is nearly 2.4 times the time constant τ, 2.4τ. The response process is quick without oscillation. The response waveform in time-domain is shown in Figure 8A.1 with $\tau_d = 0.25\tau$.

The impulse interference response is

$$\Delta g(t) = \frac{4U}{\tau} te^{-\frac{2t}{\tau}} \tag{8A.10}$$

where U is the interference signal. The transient process is still nearly 2.4 times the time constant, 2.4τ. The response waveform in time-domain is shown in Figure 19A.2 with $\tau_d = 0.25\tau$.

A4 Large Damping Time Constant

If the damping time constant is large (i.e. $\tau_d > \tau/4$, $\xi > 0.25$), the transfer function (8A.1) is a second-order function with a couple of conjugated complex poles, $-s_1$ and $-s_2$, in the left-hand half plane (LHHP) in s-domain

$$G(s) = \frac{M}{1 + s\tau + s^2\tau\tau_d} = \frac{M / \tau\tau_d}{(s + s_1)(s + s_2)} \tag{8A.11}$$

where

$$s_1 = \sigma + j\omega \quad \text{and} \quad s_2 = \sigma - j\omega$$

$$\sigma_1 = \frac{1}{2\tau_d} \quad \text{and} \quad \omega = \frac{\sqrt{4\tau_d - \tau^2}}{2\tau\tau_d}$$

There are a couple of conjugated complex poles, $-s_1$ and $-s_2$, in the transfer function. This expression describes the characteristics of the DC/DC converter. The unit-step function response in the time-domain is

$$g(t) = M\left[1 - e^{-\frac{t}{2\tau_d}}\left(\cos\omega t - \frac{1}{\sqrt{4\tau_d / \tau - 1}}\sin\omega t\right)\right] \tag{8A.12}$$

The transient response has oscillation progress with damping factor σ and frequency ω. The corresponding waveform in time-domain is shown in Figure 8A.1 with $\tau_d = 0.5\tau$, and in Figure 8A.3 with τ, 2τ,5τ and 10τ.

The impulse interference response is

$$\Delta g(t) = \frac{U}{\sqrt{\frac{\tau_d}{\tau} - \frac{1}{4}}} e^{-\frac{t}{2\tau_d}} \sin(\omega t) \tag{8A.13}$$

where U is the interference signal. The recovery process is a curve with damping factor σ and frequency ω. The response waveform in time-domain is shown in Figure 8A.2 with $\tau_d = 0.5\tau$, and in Figure 8A.4 with τ, 2τ,5τ and 10τ.

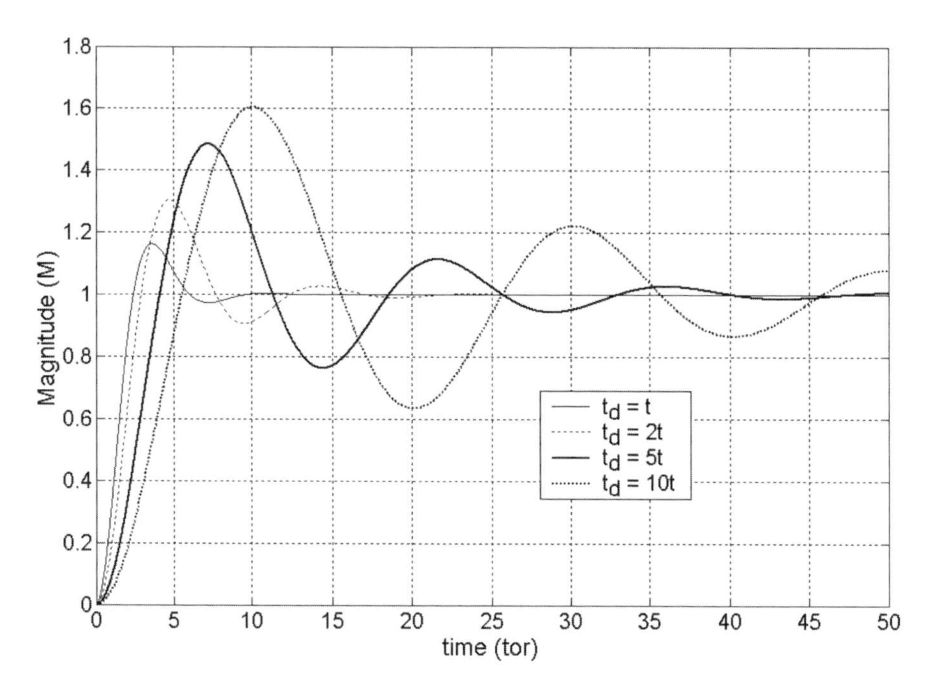

FIGURE 8A.3
Unit-step function responses ($\tau_d = \tau$, 2τ, 5τ and 10τ).

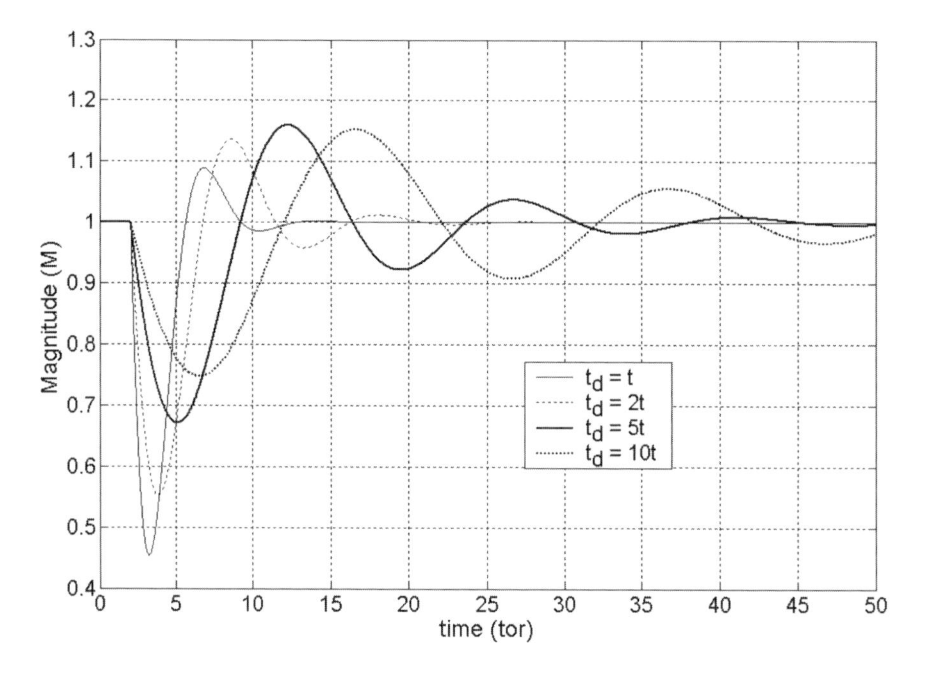

FIGURE 8A.4
Impulse responses ($\tau_d = \tau$, 2τ, 5τ and 10τ).

Appendix B: Some Calculation Formulas Derivations

B1 Transfer Function of Buck Converter

Referring to Figure 8.3, we obtain the output voltage $v_2(t)$ from input source voltage $v_1(t) = V_1$ using the voltage division formula.

$$v_1(t) = \begin{cases} V_1 & 0 \le t < kT \\ 0 & kT \le t < T \end{cases}$$

Correspondingly, the transfer function in the s-domain is

$$G(s) = \frac{v_2(s)}{v_1(s)} = M \frac{\dfrac{R\dfrac{1}{sC}}{R+\dfrac{1}{sC}}}{sL+\dfrac{R\dfrac{1}{sC}}{R+\dfrac{1}{sC}}} = \frac{M}{1+s\dfrac{L}{R}+s^2LC} \tag{8B.1}$$

where M is the voltage transfer gain in the steady state.

This transfer function is in the second-order form. It is available for other fundamental converters consisting of two passive energy-storage elements (one inductor L and one capacitor C) and load R such as boost converter and buck-boost converter. The voltage transfer gain $M = k$ for buck converter, $M = 1/(1 - k)$ for boost converter, $M = k/(1 - k)$ for buck-boost converter.

B2 Transfer Function of Super-Lift Luo-Converter

Refer to Figure 8.16, we obtain the output voltage $v_2(t)$ from input source voltage $v_1(t) = V_1$ $(0\ \ t < kT)$ using the voltage division formula. Correspondingly, the transfer function in the s-domain is

$$G(s)=\frac{v_2(s)}{v_1(s)}=M\frac{\dfrac{R\dfrac{1}{sC_2}}{R+\dfrac{1}{sC_2}}}{sL+\dfrac{1}{sC_1}+\dfrac{R\dfrac{1}{sC_2}}{R+\dfrac{1}{sC_2}}}=\frac{MsRC_1}{1+sR(C_1+C_2)+s^2LC_1+s^3RLC_1C_2} \tag{8B.2}$$

where M is the voltage transfer gain in the steady state. $M = (2 - k)/(1 - k)$ for the elementary circuit of a positive-output super-lift Luo-converter. This is a third-order transfer function in the s-domain.

B3 Simplified Transfer Function of Super-Lift Luo-Converter

Refering to Figure 8.16, we obtain the output voltage $v_2(t)$ from input source voltage $v_1(t) = V_1$ using the voltage division formula. Correspondingly, the transfer function in the s-domain is shown in equation (8B.2). If the capacitance C_2 is very small, it can be ignored; and the item involving the C_2 can be deleted in (8B.2). Therefore, we obtain the simplified transfer function (8B.3) below

$$G(s)=\frac{MsRC_1}{1+sRC_1+s^2LC_1} \tag{8B.3}$$

This is a second-order transfer function in the s-domain with two "poles" and one "zero," which means that there is off-set from the beginning.

In the other case, if C_1 is very large and $1/SC_1=0$, we may obtain the other from

$$G(s)=\frac{M}{1+s\dfrac{L}{R}+s^2LC_2} \tag{8B.4}$$

This is a second-order transfer function in the s-domain with two "poles." There is no off-set from the beginning.

B4 Time Constants τ and τ_d, and Ratio ξ

The deviation of time constants τ and τ_d, and ratio ξ can be referred to as the transfer function of the buck converter with power losses ($r_L \neq 0$).

$$G(s)=\frac{M}{1+\frac{r}{R}+s\left(Cr+\frac{L}{R}\right)+s^2LC}=\frac{M\eta}{1+s\left(Cr+\frac{L}{R}\right)\eta+s^2LC\eta} \tag{8B.5}$$

$$\begin{aligned}\tau&=\eta\left(Cr+\frac{L}{R}\right)=\frac{CRr}{R+r}+\eta\frac{L}{R}=(1-\eta)\frac{2W_C}{P_O}+\eta\frac{2W_L}{P_O}=\\&=\frac{2T(1-\eta)EF*CIR}{\eta(1+CIR)}+\frac{2T*EF}{1+CIR}=\frac{2T*EF}{1+CIR}\left(1+CIR\frac{1-\eta}{\eta}\right)\end{aligned} \tag{8B.6}$$

$$\begin{aligned}\tau_d&=\frac{CL\eta}{\mu\left(Cr+\frac{L}{R}\right)}=\frac{CL\eta}{\frac{2T*EF}{1+CIR}\left(1+CIR\frac{1-\eta}{\eta}\right)}=\frac{\left(\frac{2T*EF}{1+CIR}\right)^2\frac{CIR}{\eta}}{\frac{2T*EF}{1+CIR}\left(1+CIR\frac{1-\eta}{\eta}\right)}=\\&=\frac{2T*EF}{1+CIR}\frac{CIR}{\eta+CIR(1-\eta)}\end{aligned} \tag{8B.7}$$

$$\xi=\frac{\tau_d}{\tau}=\frac{CIR}{\eta\left(1+CIR\frac{1-\eta}{\eta}\right)^2} \tag{8B.8}$$

From equations (8.23) and (8B.6)

$$\begin{aligned}\tau&=\frac{2T\times EF}{1+CIR}\left(1+CIR\frac{1-\eta}{\eta}\right)=\frac{2T\times\frac{SE}{PE}}{1+CIR}\left(1+CIR\frac{1-\eta}{\eta}\right)=\\&=\frac{2\times SE/P_{in}}{1+CIR}\left(1+CIR\frac{1-\eta}{\eta}\right)\end{aligned}$$

The stored energy *SE*, *CIR*, input power P_{in} and the efficiency η are dependent on the working state, but independent from the switching frequency f and conduction duty cycle k. Therefore the time constant τ is independent from the switching frequency f and the conduction duty cycle k.

From equations (8.25) and (8B.7)

$$\tau_d = \frac{2T \times EF}{1 + CIR} \frac{CIR}{\eta + CIR(1-\eta)} = \frac{2T \times \dfrac{SE}{PE}}{1 + CIR} \frac{CIR}{\eta + CIR(1-\eta)} =$$

$$= \frac{2 \times SE / P_{in}}{1 + CIR} \frac{CIR}{\eta + CIR(1-\eta)}$$

Analogously, the time constant ratio ξ are independent from the switching frequency f and conduction duty cycle k.

Usually the stored energy is proportional to the input power. Therefore, when the working state changes from one steady state to a new one; the time constant τ, the damping time constant τ_d, and the time constant ratio ξ are unchanged. They are the parameters to rely on in circuit structure and power losses. Readers can try changing the k and/or f to repeat the exercises in Section 5, and will find the time constant τ, the damping time constant τ_d, and the time constant ratio ξ are unchanged.

Index

A

AC/DC rectifiers, *see* Energy sources, DC
Active clamped synchronous rectifier Luo-converters, 7–9
Attenuation rate, two-element RPC inductor current, 36

B

Back EMF load, 176
diode rectifiers
single-phase bridge, 129–131
three-phase half-bridge, 134–136
thyristor rectifiers, single-phase half-wave with back EMF
plus inductive load, 144–145
plus pure inductor, 145–147
plus resistive load, 142–144
Bipolar current source, 29–30
single-voltage source circuits, 29–30
two-voltage source circuits, 29–30
Bipolar square wave current, Π–CLL current source resonant inverter, 43
Bipolar voltage source, 26–29
cascade reverse double Γ–LC resonant power converter, 76–77
single-voltage source circuits, 27–29
two-voltage source circuits, 26–27
Bode plot
closed loop control system diagram, 99, 100, 101, 102
open loop system, 102, 103
Boost converter in CCM, 201, 203–205
Bridge diode rectifiers, single-phase, 125–133
back EMF load, 129–131
capacitative load, 131–133
resistive load, 127–129
Buck converters
in continuous conduction mode (CCM), 186–199
with large energy losses, 196–198, 199
with moderate energy losses, 192, 194–196
with small energy losses, 189–192, 193, 194
without energy losses, 186–189, 190
small-signal analysis
with small energy losses, 215–218, 219, 220
with zero energy losses, 214–215
transfer functions, derivations of formulae, 231
Buck-boost converter in CCM, 206–208

C

Capacitors
cascade double Γ–CL current source resonant inverter components, 59
stored energy ratios, capacitor-inductor (CIR), 178
stored energy variation on (VE), 179–180
Cascade double Γ–CL current source resonant inverter, 57–73
applying frequency, 73
current transfer gain < 1, 73
experimental results, 71–73
function of double Γ–CL circuit, 73
mathematical analysis, 57–67
components, voltages and currents, 59–60
input impedance, 58–59
power transfer efficiency, 66–67
simplified impedance and current gain, 60–66
simulation results, 67–71
Cascade reverse double Γ–LC resonant power converter, 75–113, 114
classical analysis, AC side, 77–84
equivalent AC circuit and transfer functions, 78–80
equivalent load resistance, 77–78
operating principles, 77

voltage transfer gain and input impedance, 80–84
closed-loop system design, 99–104, 105
discontinuous conduction mode (DCM), 108–113, 114
experimental results, 86
resonance operation and modeling, 86–92
operating principles, operating modes, and equivalent circuits, 87–89
state-space analysis, 89–92
simulation results, 85
small-signal modeling, 93–99
diagram of model, 93
equivalent circuit model, 98–99
extended describing function, 95–96
harmonic approximation, 94–95
harmonic balance, 96–97
nonlinear state equation, 93–94
parameters used in, 116
perturbation and linearization, 97–98
steady-state analysis, 76–86
topology and circuit description, 76–77
variable-parameter resonant converter characteristics, 105–108
Classical analysis, cascade reverse double Γ–LC resonant power converter, 77–84
equivalent AC circuit and transfer functions, 78–80
equivalent load resistance, 77–78
operating principles, 77
voltage transfer gain and input impedance, 80–84
Closed-loop system design, cascade reverse double Γ–LC resonant power converter, 99–104, 105
Compensation system, cascade reverse double Γ–LC resonant power converter, 99–103
Conduction mode, continuous, *see* Continuous conduction mode (CCM)
Conduction mode, discontinuous
cascade reverse double Γ–LC resonant power converter, 108–113, 114
stored energy in, 180–183
Continuous conduction mode (CCM)
buck converter in, 186–199; *see also* Stored energy in continuous conduction mode (CCM)
with large energy losses, 196–198, 199
with moderate energy losses, 192, 194–196
with small energy losses, 189–192, 193, 194
without energy losses, 186–189, 190
cascade reverse double Γ–LC resonant power converter, 86–89
stored energy in, *see* Stored energy in continuous conduction mode (CCM)
Control circuit, 157–173
applications, DC/DC converters, 168–173
IBM 1.8 v/200A power supply, 171–173
insulation test bench, 5000 V., 168–169
MIT 42/24 V 3KW converter, 169–171
EMI, EMS, and EMC, 161–168
comparison with hard- and soft-switching, 153
EMI/EMC analysis, 161–162
measuring method and results, 163–167
minimization of EMI/EMC, designing rules for, 167
Luo-resonator, 157–161
calculation formulae, 159–160
circuit, 158–159
design example, 160
Conversion efficiency
cascade reverse double Γ–LC resonant power converter, 75
switched-mode power converter, 75
Crossover frequency, unity-gain, 101, 102
Current
cascade double Γ–CL current source resonant inverter, 59–60
Π–CLL current source resonant inverter components, 44–45
Current gain, *see also* Current transfer gain
cascade double Γ–CL current source resonant inverter, 60–66
Π–CLL current source resonant inverter, 45–52
Current source, Π–CLL current source resonant inverter, 41, 42
Current source resonant inverter (CSRI), *see* Cascade double Γ–CL current source resonant inverter; Π–CLL current source resonant inverter
Current switching, zero-current-switching Luo-converter, 12–14
Current transfer gain
cascade double Γ–CL current source resonant inverter
components, voltages and currents, 60
$g < 1$, 73

simplified impedance and current gain, 61–65
simulation results, 69–70
cascade reverse double Γ–LC resonant power converter, unity-gain crossover frequency, 101
Π–CLL current source resonant inverter, 46–48, 55
two-element RPCs, 32–33
Current waveforms, input
cascade double Γ–CL current source resonant inverter, 66
two-element RPC analysis, 37, 38

D

Damping factor, two-element RPC inductor current, 36
Damping ratio, two-element RPC inductor current, 36
Damping time constant, *see* Time constant, damping time constant, and ratio
DC energy sources for converters, *see* Energy sources, DC
DC/DC converters, synchronous, *see* Synchronous rectifier DC/DC converters
Demagnetizing process, synchronous rectifier DC/DC converters,
active clamped Luo-converter, 8–9
double current Luo-converter, 11
flat transformer Luo-converter, 6
zero-current-switching Luo-converter, 13–14
zero-voltage-switching Luo-converter, 16
Diode rectifiers
single-phase bridge, 125–133
back EMF load, 129–131
capacitative load, 131–133
resistive load, 127–129
single-phase half-wave, 118–125
back EMF plus inductor load, 125
back EMF plus resistor load, 123–124
inductive load, 119–121, 122
pure inductive load, 122–123
resistive load, 118–119
synchronous rectifier DC/DC converters, 1, 2
three-phase full-bridge with resistive load, 136–138
three-phase half-bridge, 133–136
back EMF load, 134–136
resistive load, 133–134
Diodes, Schottky, 1, 2–3
Discontinuous conduction mode (DCM)
cascade reverse double Γ–LC resonant power converter, 108–113, 114
stored energy in, 180–183
Double current synchronous rectifier Luo-converters, 9–12
Dynamic clamp circuit, 1–2

E

Efficiency, Π–CLL current source resonant inverter, 55
Efficiency gain, forward converters, 1–2
Electromagnetic interference, susceptibility, and compatibility (EMI, EMS, and EMC)
cascade reverse double Γ–LC resonant power converter, 75
control circuit, 161–168
comparison with hard- and soft-switching, 163
EMI/EMC analysis, 161–162
measuring method and results, 163–167
minimization of EMI/EMC, designing rules for, 167
EMF load, 176
diode rectifiers
single-phase bridge, 129–131
three-phase half-bridge, 134–136
thyristor rectifiers, single-phase half-wave
with back EMF plus inductive load, 144–145
with back EMF plus pure inductor, 145–147
with back EMF plus resistive load, 142–144
Energy factor (EF), 175–234
applications, 186–212
boost converter in CCM, 201, 203–205
buck-boost converter in CCM, 206–208
positive output Luo-converter in CCM, 209–211, 212
super-lift Luo-converter in CCM, 198–201, 202, 203
applications, buck converter in CCM, 186–199
with large energy losses, 196–198, 199
with moderate energy losses, 192, 194–196

with small energy losses, 189–192, 193, 194
without energy losses, 186–189, 190
damping time constant, 184
energy factor definition, 182–183
formulae, derivations of, 231–234
simplified transfer function of super-lift Luo-converter, 232
time constants, damping time constants, and ratios, 232–234
transfer function of buck converter, 231
transfer function of super-lift Luo-converter, 231–232
mathematical modeling for power DC/DC converters, 185–186
pumping energy, 177
second-order transfer functions, 225–230
critical damping time constant, 228
large damping time constant, 228–229, 230
small damping time constant, 226–227
very small damping time constant, 225, 226
small-signal analysis, 211–221, 222, 223
buck converter with small energy losses, 215–218, 219, 220
buck converter without energy losses, 214–215
super-lift Luo-converter with energy losses, 218–221, 222
stored energy in continuous conduction mode (CCM), 177–180
capacitor-inductor stored energy ratio (CIR), 178
energy losses, 179
stored energy, 178
stored energy variation on inductors and capacitors, 179–180
stored energy in discontinuous conduction mode (DCM), 180–183
time constant, 183–184
time constant ratios, 184–185
variation energy factor, 183
Energy sources, DC, 117–155
single-phase bridge diode rectifier, 125–133
back EMF load, 129–131
capacitative load, 131–133
resistive load, 127–129
single-phase half-wave diode rectifier, 118–125
back EMF plus inductor load, 125
back EMF plus resistor load, 123–124
inductive load, 119–121, 122
pure inductive load, 122–123
resistive load, 118–119
three-phase full-bridge diode rectifier with resistive load, 136–138
three-phase half-bridge diode rectifier, 133–136
back EMF load, 134–136
resistive load, 133–134
thyristor rectifiers, single-phase, 138–149
full-controlled with inductive load, 148–149
full-wave semicontrolled rectifier with inductive load, 147–148
half-wave with back EMF plus inductive load, 144–145
half-wave with back EMF plus pure inductor, 145–147
half-wave with back EMF plus resistive load, 142–144
half-wave with inductive load, 140–141
half-wave with pure inductive load, 141–142
half-wave with resistive load, 139–146
thyristor rectifiers, three-phase, 149–155
full-wave with inductive load, 153–155
full-wave with resistive load, 152–153
half-wave with inductive load, 151–152
half-wave with resistive load, 149–151
Equivalent AC circuit and transfer functions, cascade reverse double Γ–LC resonant power converter, 78–80
Equivalent circuit model
cascade reverse double Γ–LC resonant power converter small-signal modeling, 98–99
two-element RPC analysis, 31
Equivalent load resistance, cascade reverse double Γ–LC resonant power converter, 77–78
Equivalent series resistor (ESR), 102
Extended describing function, cascade reverse double Γ–LC resonant power converter small-signal modeling, 95–96

F

Fast Fourier transform (FFT) spectra

cascade double Γ–CL current source resonant inverter, 66, 68, 69
two-element RPC inductor current, 38
Filters, 176
cascade double Γ–CL current source resonant inverter, 66
cascade reverse double Γ–LC resonant power converter, 77
Π–CLL current source resonant inverter, 54
resonant power converters, 24–25
two-element RPC analysis, 31
Flat transformer synchronous rectifier Luo-converters, 5–7
Forward converters, 1–4
Forward synchronous rectifiers (FSRs), 3–4
Four-element current source resonant inverter, *see* Cascade double Γ–CL current source resonant inverter; Cascade reverse double Γ–LC resonant power converter
Four-element resonant power converters, 22–24
Frequency
cascade double Γ–CL current source resonant inverter, 73
Π–CLL current source resonant inverter, 55
Full-bridge diode rectifiers, three-phase with resistive load, 136–138
Full-controlled thyristor rectifier with inductive load, single-phase, 148–149
Full-wave thyristor rectifiers
single-phase, semicontrolled with inductive load, 147–148
three-phase
with inductive load, 153–155
with resistive load, 152–153

G

Gain, current transfer, *see* Current transfer gain
Gain, voltage transfer, cascade reverse double Γ–LC resonant power converter, 80–84, 105–106
Γ–CL current source resonant inverter, *see* Cascade double Γ–CL current source resonant inverter
Γ–LC current source resonant inverter, *see* Cascade reverse double Γ–LC resonant power converter

H

Half-bridge converter
cascade reverse double Γ–LC resonant power converter, 77, 99–100
three-phase diode rectifiers, 133–136
back EMF load, 134–136
resistive load, 133–134
Half-wave rectifiers
diode, single-phase, 118–125
back EMF plus inductor load, 125
back EMF plus resistor load, 123–124
inductive load, 119–121, 122
pure inductive load, 122–123
resistive load, 118–119
thyristor, single-phase
with back EMF plus inductive load, 144–145
with back EMF plus pure inductor, 145–147
with back EMF plus resistive load, 142–144
with inductive load, 140–141
with pure inductive load, 141–142
with resistive load, 139–146
thyristor, three-phase
with inductive load, 151–152
with resistive load, 149–151
Hard switching, EMI, EMS, and EMC comparisons, 163
Harmonic approximation, cascade reverse double Γ–LC resonant power converter small-signal modeling, 94–95
Harmonic balance, cascade reverse double Γ–LC resonant power converter small-signal modeling, 96–97
Harmonics, two-element RPC inductor current, 38

I

IBM 1.8 v/200A power supply, 171–173
Impedance
cascade double Γ–CL current source resonant inverter, 58–59, 60–66
input, *see* Input impedance
Π–CLL current source resonant inverter, 43–44, 45–52
Inductive load, rectifiers
diode, 119–121, 122; *see also* Diode rectifiers
thyristor, *see* Thyristor rectifiers

Inductors
cascade double Γ–CL current source resonant inverter, 59–60
Π–CLL current source resonant inverter, component voltages and currents, 44
stored energy ratios, capacitor-inductor (CIR), 178
stored energy variation on (VE), 179–180
two-element RPC analysis, 36
Input impedance
cascade double Γ–CL current source resonant inverter, 58–59, 66
cascade reverse double Γ–LC resonant power converter, 80–84
Π–CLL current source resonant inverter, 43–44
resonant power converters, two-element, 31–32
Input square wave current, two-element RPC inductor current, 37
Insulation test bench, 5000 V., 168–169
Inverters, *see* Cascade double Γ–CL current source resonant inverter; Π–CLL current source resonant inverter

K

Kirchoff's current law, cascade reverse double Γ–LC resonant power converter, 99

L

Linearization, cascade reverse double Γ–LC resonant power converter small-signal modeling, 97–98
Load, 176
back EMF, 176; *see also* Diode rectifiers; Thyristor rectifiers
Π–CLL current source resonant inverter, 42
rectifier, *see* Diode rectifiers; Thyristor rectifiers
Load resistance, cascade reverse double Γ–LC resonant power converter, classical analysis, AC side, 77–78
Luo-converters, *see also* Synchronous rectifier DC/DC converters
positive output, in CCM, 209–211, 212
small-signal analysis, 218–221, 222
transfer functions, derivations of formulae, 231–232
Luo-resonator, control circuit, 157–161
calculation formulae, 159–160
circuit explanation, 158–159
design example, 160

M

Magnetizing process, synchronous rectifier DC/DC converters,
active clamped Luo-converter, 8
double current Luo-converter, 10–11
flat transformer Luo-converter, 5–6
zero-current-switching Luo-converter, 13
zero-voltage-switching Luo-converter, 15–16
Mathematical analysis
cascade double Γ–CL current source resonant inverter, 57–67
components, voltages and currents, 59–60
input impedance, 58–59
power transfer efficiency, 66–67
simplified impedance and current gain, 60–66
cascade reverse double Γ–LC resonant power converter
classical analysis, AC side, 77–84
steady-state analysis, 76–86
Π–CLL current source resonant inverter, 43–53
impedance and current gain, simplified, 45–52
input impedance, 43–44
power transfer efficiency, 52–53
voltages and currents of components, 44–45
Mathematical modeling
cascade reverse double Γ–LC resonant power converter, 85–92
operating principles, operating modes, and equivalent circuits, 87–89
small-signal modeling, *see* Small-signal modeling
state-space analysis, 89–92
derivation of formulae, 231–234
energy factor (EF), *see also* Energy factor (EF)
second-order transfer functions, 225–230

stored energy in discontinuous conduction mode (DCM), 180–183
time constant and damping time constant, 183–185
small-signal analysis
buck converter with small energy losses, 215–218, 219, 220
buck converter without energy losses, 214–215
super-lift Luo-converter with energy losses, 218–221, 222
Matlab, 109
MIT 42/24 V 3KW converter, 169–171
MOSFETS, 1, 3, 4, 5
cascade reverse double Γ–LC resonant power converter, 76–77
Multiple energy-storage element resonant power converters, 19–38
bipolar current source, 29–30
single-voltage source circuits, 29–30
two-voltage source circuits, 29–30
bipolar voltage source, 26–29
single-voltage source circuits, 27–29
two-voltage source circuits, 26–27, 29
cascade, *see* Cascade reverse double Γ–LC resonant power converter
four-element, 22–24
Γ–CL current source resonant inverter, *see* Cascade double Γ–CL current source resonant inverter
Π–CLL current source resonant inverter, *see* Π–CLL current source resonant inverter
three-element, 21–22
two-element, 20–21
two-element RPC analysis, 31–38
current transfer gain, 32–33
experimental results, 37, 38
input impedance, 31–32
operation analysis, 33–36
simulation results, 37–38

N

Nonlinear state equation, cascade reverse double Γ–LC resonant power converter small-signal modeling, 93–94

O

Operation analysis, resonant power converters, two-element, 33–37
Oscillatory component, two-element RPC inductor current, 36
Output current gain, cascade double Γ–CL current source resonant inverter, simplified impedance and current gain, 60
Output current waveform, two-element RPC analysis, 37, 38
Output power, cascade double Γ–CL current source resonant inverter, 67

P

Perturbation, cascade reverse double Γ–LC resonant power converter small-signal modeling, 97–98
Phase lag, compensation network and, 101–103
Π–CLL current source resonant inverter, 41–55
applying frequency, 55
current source, 41, 42
DC current component, 55
efficiency, 55
function of Π–CLL circuit, 54–55
g >1, 55
load, 42
mathematical analysis, 43–53
impedance and current gain, simplified, 45–52
input impedance, 43–44
power transfer efficiency, 52–53
voltages and currents of components, 44–45
pump circuits, 41
resonant circuit, 42
simulation results, 53
Power density
cascade reverse double Γ–LC resonant power converter, 75
switched-mode power converter, 75
Power factor, 175–176
cascade reverse double Γ–LC resonant power converter, 75
switched-mode power converter, 75
Power transfer efficiency, 175–176
cascade double Γ–CL current source resonant inverter, 66–67

Π–CLL current source resonant inverter, 52–53
PSpice, 85, 99, 109
Pulse-width-modulating (PWM) converters, switching losses, 75–76
Pulse-width-modulation (PWM) signal, 3
Pump circuits, Π–CLL current source resonant inverter, 41, 42–43

Q

Q values
- cascade double Γ–CL current source resonant inverter, 63, 64, 65, 67, 68
- cascade reverse double Γ–LC resonant power converter, 105, 106, 107
- Π–CLL current source resonant inverter, 46–52

Quantization, energy, 177

R

Rectifiers, *see* Diode rectifiers; Synchronous rectifier DC/DC converters; Thyristor rectifiers
Resistance, load, cascade reverse double Γ–LC resonant power converter, 77–78
Resistive load, rectifiers
- half-wave diode, 118–119
- thyristor, *see* Thyristor rectifiers

Resistors, Π–CLL current source resonant inverter components, 45
Resonance operation and modeling, cascade reverse double Γ–LC resonant power converter, 86–92
- operating principles, operating modes, and equivalent circuits, 87–89
- state-space analysis, 89–92

Resonant circuit, Π–CLL current source resonant inverter, 42
Resonant inverters, Π–CLL current source resonant inverter, *see* Π–CLL current source resonant inverter
Resonant network, cascade reverse double Γ–LC resonant power converter, 77
Resonant period, synchronous rectifier DC/DC converters
- zero-current-switching Luo-converter, 13
- zero-voltage-switching Luo-converter, 15–16

Resonant power converters (RPC), *see* Cascade double Γ–CL current source resonant inverter; Cascade reverse double Γ–LC resonant power converter; Multiple energy-storage element resonant power converters
Resonant waveforms, resonant power converters, 76
Reverse double Γ–LC resonant power converter, *see* Cascade reverse double Γ–LC resonant power converter
Ripple factor, 175
RPC, *see* Multiple energy-storage element resonant power converters

S

Schottky diode, 1, 2–3
Single-phase rectifiers
- diode, *see* Diode rectifiers
- thyristor, *see* Thyristor rectifiers

Single-voltage source circuits, multiple energy-storage element resonant power converters
- bipolar current source, 30
- bipolar voltage source, 27–29

Sinusoidal function, two-element RPC inductor current, 36
Sinusoidal waveform
- cascade double Γ–CL current source resonant inverter, 67, 68, 69
  - experimental results, 72
  - simulation results, 70, 71
- cascade reverse double Γ–LC resonant power converter simulation studies, 85, 86
- DC energy sources for converters, *see* Energy sources, DC
- Π–CLL current source resonant inverter, 52, 54

Small-signal analysis, 211–221, 222, 223
- buck converter
  - with small energy losses, 215–218, 219, 220
  - without energy losses, 214–215
- super-lift Luo-converter with energy losses, 218–221, 222

Small-signal modeling

cascade reverse double Γ–LC resonant power converter, 93–99
diagram of model, 93
equivalent circuit model, 98–99
extended describing function, 95–96
harmonic approximation, 94–95
harmonic balance, 96–97
nonlinear state equation, 93–94
parameters used in, 116
perturbation and linearization, 97–98
Soft-switching, EMI, EMS, and EMC comparisons, 163
Square wave current
Π–CLL current source resonant inverter, 43, 54
two-element RPC inductor current, 37
Square waveform
cascade double Γ–CL current source resonant inverter, 66, 70, 71
cascade reverse double Γ–LC resonant power converter, 77, 85, 86
State equation
cascade reverse double Γ–LC resonant power converter small-signal modeling, 93–94
two-element RPC analysis, 33–34
Steady-state analysis, cascade reverse double Γ–LC resonant power converter, 76–86
Stored energy in continuous conduction mode (CCM), 177–180
capacitor-inductor stored energy ratio (CIR), 178
energy losses, 179
stored energy, 178
stored energy variation on inductors and capacitors, 179–180
Stored energy in discontinuous conduction mode (DCM), 180–182
Super-lift Luo-converter
in continuous conduction mode (CCM), 198–201, 202, 203
small-signal analysis, 218–221, 222
transfer functions, derivations of formulae, 231–232
Switched-mode power converter, specifications, 75
Switching
EMI, EMS, and EMC comparisons, 163
synchronous rectifier DC/DC converters
active clamped Luo-converter, 8–9
double current Luo-converter, 11
flat transformer Luo-converter, 6, 7
zero-current-switching Luo-converter, 13, 14
zero-voltage-switching Luo-converter, 16
Switching frequency
cascade reverse double Γ–LC resonant power converter, 77
discontinuous conduction mode (DCM), 112–113, 114
pulse-width-modulating (PWM) converters, 75–76
switched-mode power converter, 75
Switching losses, pulse-width-modulating (PWM) converters, 76
Synchronous rectifier DC/DC converters, 1–17
active clamped Luo-converter, 7–9
double current Luo-converter, 9–12
flat transformer Luo-converter, 5–7
zero-current-switching Luo-converter, 12–14
zero-voltage-switching Luo-converter, 14–17

T

Three-element resonant power converters, 21–22; *see also* Π–CLL current source resonant inverter
Three-phase rectifiers
diode, *see* Diode rectifiers
thyristor, *see* Thyristor rectifiers
Thyristor rectifiers
single-phase, 138–149
full-controlled rectifier with inductive load, 148–149
full-wave semicontrolled with inductive load, 147–148
half-wave with back EMF plus inductive load, 144–145
half-wave with back EMF plus pure inductor, 145–147
half-wave with back EMF plus resistive load, 142–144
half-wave with inductive load, 140–141
half-wave with pure inductive load, 141–142
half-wave with resistive load, 139–146
three-phase, 149–155
full-wave with inductive load, 153–155
full-wave with resistive load, 152–153
half-wave with inductive load, 151–152

half-wave with resistive load, 149–151
Time constant, damping time constant, and ratio, 183–185
derivation of formulae, 232–234
second-order transfer functions
critical, 228
large, 228–229, 230
small, 226–227
very small, 225, 226
Time constant, energy factor (EF), 183–184
Time constant ratios, 184–185
Time domain, state equation in, two-element RPC analysis, 32–33
Time-attenuation, two-element RPC inductor current, 35
Topology
cascade reverse double Γ–LC resonant power converter, 76–77
four-element RPC, 22–24
three-element RPC, 21–23
two-element RPC, 20–21
Total harmonic distortion, 175, 176
cascade reverse double Γ–LC resonant power converter, 75, 108, 110
switched-mode power converter, 75
Transfer functions
cascade reverse double Γ–LC resonant power converter
classical analysis, AC side, 78–80
compensation system, 100
open loop system, 104
formulae, derivations of, 231–234
buck converter, 231
super-lift Luo-converter, 231–232
super-lift Luo-converter, simplified, 232
Π–CLL current source resonant inverter, 44
second-order, 225–230
critical damping time constant, 228
large damping time constant, 228–229, 230
small damping time constant, 226–227
very small damping time constant, 225, 226
Transformer, synchronous rectifier converters, *see* Synchronous rectifier DC/DC converters
Two-element resonant power converters, 20–21
Two-element RPC analysis, 31–38
current transfer gain, 32–33
experimental results, 38
input impedance, 31–32
operation analysis, 33–36
simulation results, 37–38
Two-voltage source circuits
bipolar current source, 29, 30
bipolar voltage source, 26–27, 29

U

Unity-gain crossover frequency, 101, 102

V

Variable-parameter resonant converter characteristics, cascade reverse double Γ–LC resonant power converter, 105–108
Variation energy factor, 183
Voltage controlled oscillator (VCO), cascade reverse double Γ–LC resonant power converter, 99
Voltage switching, zero-voltage-switching Luo-converter, 14–17
Voltage transfer gain, cascade reverse double Γ–LC resonant power converter, 80–84, 105–106, 107–108
Voltage waveforms, two-element RPC analysis, 36
Voltages
cascade double Γ–CL current source resonant inverter, 59–60
Π–CLL current source resonant inverter components, 44–45

W

Waveforms, *see also* Sinusoidal waveform; Square waveform
cascade double Γ–CL current source resonant inverter, experimental results, 72
current, *see* Current waveforms, input
switching, 76
two-element RPC analysis, 35, 36, 37, 38

Z

Zero-current-switching synchronous rectifier
Luo-converters, 12–14
Zero-voltage-switching Luo-converter, 14–17

RECEIVED
MAR - 7 2006
ENGINEERING LIBRARY